T5-DGH-466

HOMOLOGY

The Novartis Foundation is an international scientific and educational charity (UK Registered Charity No. 313574). Known until September 1997 as the Ciba Foundation, it was established in 1947 by the CIBA company of Basle, which merged with Sandoz in 1996, to form Novartis. The Foundation operates independently in London under English trust law. It was formally opened on 22 June 1949.

The Foundation promotes the study and general knowledge of science and in particular encourages international co-operation in scientific research. To this end, it organizes internationally acclaimed meetings (typically eight symposia and allied open meetings, 15–20 discussion meetings, a public lecture and a public debate each year) and publishes eight books per year featuring the presented papers and discussions from the symposia. Although primarily an operational rather than a grant-making foundation, it awards bursaries to young scientists to attend the symposia and afterwards work for up to three months with one of the other participants.

The Foundation's headquarters at 41 Portland Place, London W1N 4BN, provide library facilities, open every weekday, to graduates in science and allied disciplines. The library is home to the Media Resource Service which offers journalists access to expertise on any scientific topic. Media relations are also strengthened by regular press conferences and book launches, and by articles prepared by the Foundation's Science Writer in Residence. The Foundation offers accommodation and meeting facilities to visiting scientists and their societies.

Information on all Foundation activities can be found at http://www.novartisfound.org.uk

Novartis Foundation Symposium 222

HOMOLOGY

1999

JOHN WILEY & SONS

Chichester · New York · Weinheim · Brisbane · Singapore · Toronto

Published in 1999 by John Wiley & Sons Ltd,
Baffins Lane, Chichester,
West Sussex PO19 1UD, England

National 01243 779777
International (+44) 1243 779777
e-mail (for orders and customer service enquiries): cs-books@wiley.co.uk
Visit our Home Page on http://www.wiley.co.uk
or http://www.wiley.com

Other Wiley Editorial Offices

John Wiley & Sons, Inc., 605 Third Avenue,
New York, NY 10158-0012, USA

WILEY-VCH Verlag GmbH, Pappelallee 3,
D-69469 Weinheim, Germany

Jacaranda Wiley Ltd, 33 Park Road, Milton,
Queensland 4064, Australia

John Wiley & Sons (Asia) Pte Ltd, 2 Clementi Loop #02-01,
Jin Xing Distripark, Singapore 129809

John Wiley & Sons (Canada) Ltd, 22 Worcester Road,
Rexdale, Ontario M9W 1L1, Canada

Novartis Foundation Symposium 222
ix+256 pages, 36 figures, 3 tables

British Library Cataloguing in Publication Data

A catalogue record for this book is available from the British Library

ISBN 0 471 98493 0

Typeset in 10½ on 12½ pt Garamond by Dobbie Typesetting Limited, Tavistock, Devon.
Printed and bound in Great Britain by Biddles Ltd, Guildford and King's Lynn.
This book is printed on acid-free paper responsibly manufactured from sustainable forestry, in which at least two trees are planted for each one used for paper production.

Contents

Symposium on Homology, held at the Novartis Foundation, London 21–23 July 1998

This symposium is based on a proposal made by Adam Wilkins

Editors: Gregory R. Bock (Organizer) and Gail Cardew

Participants

E. Abouheif Department of Ecology and Evolution, State University of New York at Stony Brook, Stony Brook, NY 11794-5245, USA

M. Akam University Museum of Zoology, Downing Street, Cambridge CB2 3EJ, UK

R. L. Carroll Redpath Museum, McGill University, 859 Sherbrooke St. West, Montreal, Canada H3A 2K6

P. Donoghue School of Earth Sciences, University of Birmingham, Edgbaston, Birmingham B15 2TT, UK

F. Galis Institute for Evolutionary and Ecological Sciences, University of Leiden, PO Box 9516, 2300 RA Leiden, The Netherlands

H.W. Greene Museum of Vertebrate Zoology and Department of Integrative Biology, University of California, Berkeley, CA 94720-3101, USA

B. K. Hall (*Chair*) Department of Biology, Dalhousie University, Halifax, Nova Scotia, Canada B3H 4J1

J. R. Hinchliffe Institute of Biological Sciences, University of Wales, Aberystwyth, Wales SY23 3DA, UK

N. D. Holland Marine Biology Research Division, Scripps Institution of Oceanography, La Jolla, CA 92093-0202, USA

P.W. H. Holland School of Animal and Microbial Sciences, The University of Reading, Whiteknights, Reading RG6 6AJ, UK

T. C. Lacalli Department of Biology, University of Saskatchewan, Saskatoon, Saskatchewan, Canada S7N 5E2

J. Maynard Smith School of Biological Sciences, University of Sussex, Falmer, Brighton BN1 9QG, UK

A. Meyer Department of Biology, University of Konstanz, 78457 Konstanz, Germany

G. B. Müller Department of Anatomy, University of Vienna, Währingerstrasse 13, A-1090 Wien, and Konrad Lorenz Institute for Evolution and Cognition Research, Adolf-Lorenz-Gasse 2, A-3422 Altenberg, Austria

A. L. Panchen Department of Marine Sciences, University of Newcastle upon Tyne, NE1 7RU, UK

M. Purnell (*Bursar*) Department of Geology, University of Leicester, Leicester LE1 7RH, UK

E. Raff Indiana Molecular Biology Institute, and Department of Biology, Indiana University, Bloomington, IN 47405, USA

R. A. Raff Indiana Molecular Biology Institute, and Department of Biology, Indiana University, Bloomington, IN 47405, USA

V. L. Roth Department of Zoology, Box 90325, Duke University, Durham, NC 27708-0325, USA

G. F. Striedter Department of Psychobiology and Center for the Neurobiology of Learning and Memory, University of California at Irvine, Irvine, CA 92697, USA

D. Tautz Zoologisches Institut, Universität München, Postfach 202136, D-80021 München, Germany

G. P. Wagner Department of Ecology and Evolutionary Biology, Yale University, New Haven, CT 06520-8106, USA

D. B. Wake Museum of Vertebrate Zoology, University of California, Berkeley, CA 94720-3160, USA

A. Wilkins Editorial Office, 10/11 Tredgold Lane, Napier Street, Cambridge CB1 1HN, UK

G. A. Wray Department of Ecology and Evolution, State University of New York, Stony Brook, NY 11794-5425, USA

Introduction

Brian K. Hall

Department of Biology, Dalhousie University, Halifax, Nova Scotia, Canada B3H 4J1

I am delighted to have the opportunity of welcoming the speakers and discussants to what should be an exciting symposium on homology. Although it is presumptuous of me to set the direction that our discussion should take, and even more presumptuous to map the route we will take or to attempt to predict our destination, I will provide some signposts at the beginning of our journey. Whether the route we take or the destination we reach will bear any resemblance to the route and destination suggested remains to be seen. I pose four questions that we are likely to encounter during our deliberations.

What is homology?

Much of the discussion about homology in the past has been mired in semantics. I hope we will not become bogged down in definitional issues but rather hope that, along with David Wake (1994), in an oft-cited review of a volume on homology, we can agree that homology is 'the central concept for *all* of biology' and move on. Despite the importance of homology as the 'hierarchical basis of comparative biology' (the subtitle of the volume edited by Hall [1994]) and its central place with evolution in biology, definitions of homology abound. In the same review, and as a result of reading the 1994 book, Wake concluded that: '. . . I found no reason to change my personal definition of homology (which is not worth repeating, since I cannot even convince students in my own lab of the correctness of my position).'

Do biologists need a definition for such a central concept or is it sufficient to know that homology exists and to move on? It is my hope that in this symposium we will move on, each presenting our definition of homology for clarity of discussion, but then moving to the important issues raised by the three other issues posed below. I am sure that definitional issues will arise, but I hope they will not dominate or overwhelm us. From various speakers we will see homology as sameness, identity, synapomorphy, a statement about final structures, a statement about developmental processes and/or about genes, and we will discuss whether homology is (or should be) hierarchical and is (or can be) partial.

Perhaps by the end we may be able to agree, or at the very least comment on:

(1) whether having a definition is important or whether the explanatory power of an ill-defined but central concept is sufficient;
(2) whether one definition holds for all situations, or whether it is sufficient to just say what we each mean when we invoke homology; and
(3) whether multiple definitions are possible, allowable or desirable.

What is the antithesis of homology?

Curiously, given that homology is such a central concept, second only to evolution, non-homology is not the antithesis of homology. Traditionally, homology is contrasted with analogy, as Alec Panchen will discuss in the opening presentation. Both Panchen and Wake will raise the issue of whether homoplasy is a more appropriate antithesis to homology. Homoplasy has variously been viewed as non-homology, a mistake in homology assessment, or the occurrence of similar structures in distantly related organisms. None of these definitions coincides with analogy. I expect that our discussions will bear much more strongly on homoplasy than on analogy than has been the case in past discussions of homology.

What are the levels of homology?

A major reason for Adam Wilkins proposing homology as a topic for a Novartis Foundation Symposium was to discuss homology as a hierarchical concept. The recent finding in genetics in which equivalent (homologous) master genes 'switch on' the development of structures long regarded as non-homologous — the eyes of flies and frogs, for example — has focused attention squarely on the issue of the level(s) of biological organization at which homology should be addressed. We will discuss homology on at least three levels — structural, genetic and behavioural.

Structural homology

The basis for structural homology will be addressed directly by several speakers (Wagner, Striedter, Müller) and indirectly by many others. Is structural homology:

(1) related to some aspect of embryonic development that may be embodied in rules of development, developmental programmes or developmental constraints; or
(2) related to processes operating at one or more levels of biological hierarchy and involving constraints, epigenetic interactions or networks?

How does structural homology deal with evolutionary changes in development that may obscure homology without rendering the resulting structures non-homologous?

Genetic homology

The basis for genetic homology will be addressed in several contexts and I am sure will generate much discussion. Is genetic homology based on:

(1) the same genes;
(2) the same genes in different animals (discussed by Holland in the context of paralogy and orthology);
(3) the same genes that share the same function;
(4) the same genes with different functions (discussed by Wray); and/or
(5) similar genetic networks (discussed by Abouheif)?

Does the identification of master genes (control, developmental or regulatory genes) require a reassessment of homology?

Behavioural homology

Homology has predominantly been used for structures; hence structural homology. It would be more inclusive to speak of phenotypic homology to reflect the fact that other aspects of the phenotype, such as homologous behaviours, can be recognized and analysed. The inclusion of behavioural homology in the programme should serve to remind us that homology is much broader than structural and genetic homology. Homology is the hierarchical basis of comparative biology and that means biology at all levels. The relationship of shared or divergent structural bases to behavioural homology is similar to, and raises the same issues as, the relationship between shared or divergent embryonic development and structural homology. Just as homologous structures can arise from divergent developmental programmes, so homologous behaviours need not share a common structural basis.

What is the research agenda?

It is clear from the topics to be covered in this symposium that attention is shifting from arguments over the definition of homology to research agenda that shed light on the essential nature of homology. Numerous speakers will address research agenda in an exciting variety of contexts. These include fossils (Carroll), limb and vertebral column evolution (Galis, Hinchliffe), the nervous system (Striedter), constraints (Wagner), the origin of novelties (Müller), the evolution of early development (Raff) and behaviour (Greene). I predict that the identification of precisely how homology translates into research programmes in comparative biology—how homology is used and tested—and how research agenda shed light on homology as a central concept in biology, will be an important contribution of this symposium and this volume.

References

Hall BK (ed) 1994 Homology: the hierarchical basis of comparative biology. Academic Press, San Diego, CA

Wake DB 1994 Comparative terminology. Science 265:268–269

Homology — history of a concept

Alec L. Panchen

Department of Marine Sciences, University of Newcastle upon Tyne, NE1 7RU, UK

Abstract. The concept of homology is traceable to Aristotle, but Belon's comparison in 1555 of a human skeleton with that of a bird expressed it overtly. Before the late 18th century, the dominant view of the pattern of organisms was the *scala naturae* — even Linnaeus with his divergent hierarchical classification did not necessarily see the resulting taxonomic pattern as a natural phenomenon. The divergent hierarchy, rather than the acceptance of phylogeny, was the necessary spur to discussion of homology and the concept of analogy. Lamarck, despite his proposal of evolution, attributed homology to his *escalator naturae* and analogy to convergent acquired characters. Significantly, it was the concept of serial homology that emerged at the end of the 18th century, although comparison between organisms became popular soon after, and was boosted by the famous Cuvier/Geoffroy Saint-Hilaire debate of the 1830s. The concepts of homology and analogy were well understood by the pre- (or anti-) evolutionary comparative anatomists before the general acceptance of phylogeny, and they were defined by Owen in 1843. The acceptance of evolution led to the idea that homology should be defined by common ancestry, and to the confusion between definition and explanation. The term 'homoplasy', introduced by Lankester in 1870, also arose from a phylogenetic explanation of homology.

1999 Homology. Wiley, Chichester (Novartis Foundation Symposium 222) p 5–23

In the novel *Moby-Dick* Herman Melville (1851) frequently interrupts the narrative of the hunt for the great white whale with assorted facts, anecdotes and soliloquies. Chapter XXXII is devoted to 'cetology', including the classification of whales. Melville is in no doubt as to their taxonomic position: despite their mammalian features, they are fish: '*a whale is a spouting fish with a horizontal tail*' (his italics). He then goes on to give a classification of known whales by size (as 'Folios', 'Octavos' and 'Duodecimos'!) as all other known taxonomic characters seem to him inconsistent.

Exactly 300 years before, Pierre Belon (1551) also concluded after detailed study that whales were fish, having seen their heart, lungs, mammary glands etc., and compared them with those of humans. Yet Aristotle had described whales as mammals (Vivipara) in the fourth century B.C. (Aristotle 1945).

One's first reaction to Belon's placement of cetaceans among the fish (and also bats among the birds and the hippopotamus as a cetacean!) is either that he was ignorant of the comparative anatomy involved or perverse to the point of stupidity (Cole 1944). Yet it was also Belon (1555) who produced the famous comparative diagrams of the skeleton of a human and that of a bird, with most of the individual bones correctly homologized (Fig. 1). He claims to have dissected over 200 birds, but he also enumerated all the mammalian features of bats, and did comparative dissections of three species of cetacean. Presumably what was missing in Belon (and in Melville) was not knowledge or intelligence, but a concept of natural classification based on anything other than function and habitat — truly analogy without homology. And yet his human–bird comparison demonstrates the concept of unity of plan which today we always consider essential to the idea of homology.

Most modern biologists associate homology with the fact of evolution. Homologous structures in two different organisms are so recognized because it is proposed that the nearest common ancestor of both had a corresponding structure: indeed definitions of homology frequently involve community of descent. Historically this lacks any backing. The concept of homology was accepted by most comparative anatomists in the first half of the 19th century, based on the idea that there is a natural order of organisms whether explained by evolution or not. Classifications are to be discovered rather than invented.

I suggest that there are in fact four logical (if not strictly chronological) stages in the development of the anatomical concept of homology — other concepts, developmental, molecular and genetic are dealt with later in this book. In the first stage there can be considerable anatomical knowledge, but classification carries with it no necessary idea of a natural order. As with Aristotle and Belon homology and analogy may be recognized, but either, or a mixture of both, may be used as a basis for taxonomy. A fish is what you find on a fishmonger's slab!

The second stage is the development of criteria for the recognition of homology. Here Belon (1555) may also be invoked, but it is not until the end of the 18th century that the criteria were clearly codified. The French transcendental morphologist Etienne Geoffroy Saint-Hilaire (1772–1844) developed such criteria principally in the first volume of his *Philosophie anatomique* (Geoffroy Saint-Hilaire 1818). He noted the criterion of similarity, but his main contribution was the *principe des connexions*. Homologous structures were to be recognized by shared topology, their relationship to surrounding anatomical features.

The third stage is the recognition (hypothesis?) that there is a true natural order, whatever form that may take. Correctly identified homologies are evidence for a particular order, but the *a priori* assumption of a natural order 'out there' implies the possibility of mistaken homologies detected by their non-congruence with the taxonomic pattern produced in the light of other homologies (Patterson 1982). I consider theories as to the nature of this pattern to be of great historical importance

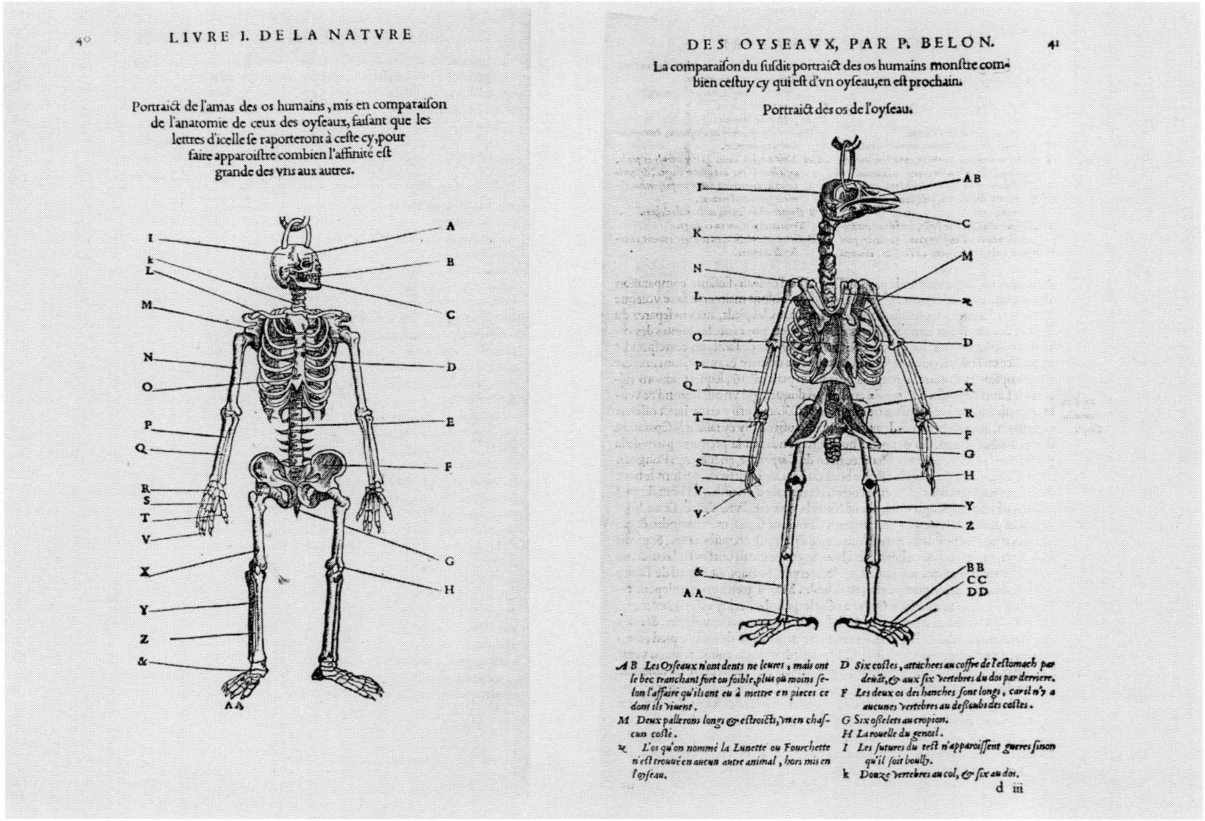

40 LIVRE I. DE LA NATVRE

Portraict de l'amas des os humains, mis en comparaison de l'anatomie de ceux des oyseaux, faisant que les lettres d'icelle se raporteront à ceste cy, pour faire apparoistre combien l'affinité est grande des vns aux autres.

DES OYSEAVX, PAR P. BELON. 41

La comparaison du susdit portraict des os humains monstre combien cestuy cy qui est d'vn oyseau, en est prochain.

Portraict des os de l'oyseau.

AB Les Oyseaux n'ont dents ne leures, mais ont le bec tranchant fort ou foible, plus ou moins selon l'affaire qu'ils ont eu à mettre en pieces ce dont ils viuent.

M Deux pallerons longs & estroicts, vn en chascun costé.

ӕ L'os qu'on nommé la Lunette ou Fourchette n'est trouvé en aucun autre animal, hors mis en l'oyseau.

D Six costes, attachees au coffre de l'estomach par deuāt, & aux six vertebres du dos par derriere.

F Les deux os des hanches sont longs, car il n'y a aucunes vertebres au dessoubs des costes.

G Six osselets au cropion.

H La rouelle du genoil.

I Les sutures du test n'apparoissent gueres sinon qu'il soit boully.

k Douze vertebres au col, & six au dos.

d iii

FIG. 1. Skeletons of a human and a bird to show homologous bones. After Belon (1551) from the original.

and will devote the next section to it. But meanwhile the fourth stage is concerned with 'mistakes'.

After the publication of *On the origin of species*. . . (Darwin 1859), systematic biologists were anxious to forge the link between the recognition of homology and the phylogenetic pattern of an irregular tree-like divergent hierarchy emphasized by both Wallace (1855) and Darwin. One of the most enthusiastic advocates of this link was Ray Lankester, and in 1870 he introduced the concept of 'homoplasy' for apparent homology not explicable by common ancestry. I will discuss this further below.

The natural order

The received opinion among most biologists is that Linnaeus inaugurated modern biological classification. This view is enhanced by the fact that the year of publication of the 10th edition of his *Systema naturae* (Linnaeus 1758) is taken as the 'starting point' by the International Code of Zoological Nomenclature (Ride et al 1985). Only taxonomic names proposed in or after that year are deemed valid. Similarly the 'start date' for the nomenclature of extant Spermophyta (flowering plants and gymnosperms), plus Pteridophyta, is based on Linnaeus' *Species plantarum* (1753). Furthermore, Linnaeus codified a series of categories, each of a different rank, so that any classification using his methods would result in a divergent hierarchy, with the animal kingdom divided into a number of classes, each class into several orders, orders into genera, and each genus into a number of species. Other intermediate categories were of course added later. Given the resulting hierarchy, homology could be defined as a sharing of features that served to diagnose a taxon at any rank, by their presence in all the species that composed that taxon (taxic homology). But if the concept of homology was (and is) to be regarded as anything other than a convenient instrumental one, the hierarchy must have some reality.

There is evidence that Linnaeus, and probably most of his naturalist contemporaries, did not regard even an imagined perfect classification as representing a natural phenomenon. Through the early part of his career, Linnaeus had been an exponent of the *scala naturae*, the natural relationships of organisms was as a ladder of perfection or elaboration from inanimate objects at the bottom to humankind at the top (Ritterbush 1964). He tried in vain to find a natural linear way to order plants, and compromised by producing an avowedly artificial hierarchical classification, his famous 'sexual system' using flower anatomy. Eventually he abandoned the search. Even in 1751 he had suggested that 'All plants show affinities on all sides, like the territories in a geographical map.' He had also suggested the analogy of a map in a letter of Johann Matthias Gesner (1691–1761: not to be confused with the great naturalist Konrad von

Gesner, 1516–1565). The map analogy was to be presented pictorially by a German pupil of Linnaeus, Paul Giseke, who edited a posthumous book by his master (Linnaeus 1792) and added a pictorial representation of his own (Fig. 2). Oldroyd (1980) has suggested a comparison between Giseke's figure and a phylogeny sliced through at a time plane, but it seems much more probable to me that a better analogy is the clustering in hyperspace of a modern phenetic taxonomist. The latter implies no hypothesis of homology.

But how could Linnaeus thirst after a *scala naturae* but propose a method of classification that yielded a divergent hierarchy? The answer was that, to him, elucidating the *scala* was a different operation from classification. His method of classification was based on the method of logical division attributed to Plato and Aristotle (Cain 1958, Panchen 1992, chapter 6). A taxon at any level (e.g. species) was characterized by features of the taxon at one rank above (genus) plus features—each a *diafora* (Greek) or *differentia* (Latin)—diagnostic of it: logical division *per genus et differentiam*. To Plato and Aristotle a *genos* (genus) could be a class at any level, from all created things to a small group, animate or not, and an *eidos* (species) one of the subordinate classes into which the *genos* was divided. Linnaeus gave genus and species an appointed rank, but as with the Greeks, his method was to systematize knowledge rather than to reveal the natural order. If there was any sense in which Linnaeus regarded his taxonomic hierarchy as natural, it is that corresponding to the philosophy of essentialism (Panchen 1992 [p 109–117, 337–341], 1994). Any taxon has a reality because its essence is represented by an exclusive list of characters. I have argued that this is to be distinguished from the Idealism of Plato, in which any valid taxon owes its reality to the existence of a Platonic '*form*' or '*idea*' corresponding to it. In its final state, in the *Republic* and the *Timaeus*, Plato was to claim not only that his 'ideas' were real, but that they were the only real entities (Oldroyd 1986, chapter 1). But before we turn to idealistic morphology, there is more that needs to be said about the *scala naturae*.

The curse of the *scala naturae*

In his famous series of William James lectures on the *scala naturae*, published as '*The great chain of being*', Arthur O. Lovejoy (1936, chapter V) quotes from a letter written by the philosopher Gottfried Wilhelm Leibniz (1646–1716) and published in 1753. In part this reads:

> . . . all the orders of natural beings form but a single chain, in which the various classes, like so many rings, are so closely linked to one another that it is impossible for the senses or the imagination to determine precisely the point at which one ends and the next begins—all the species which, so to say, lie

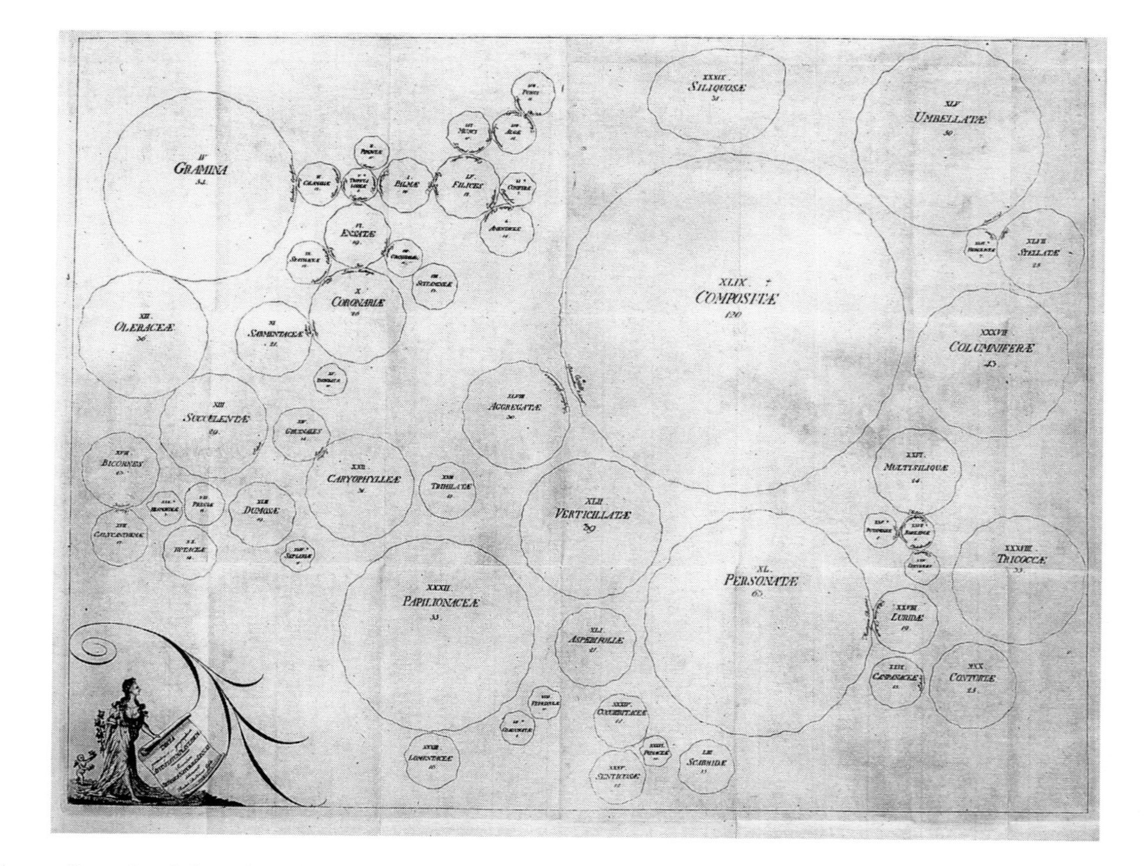

FIG. 2. A two-dimensional clustering of plant families by P. D. Giseke in Linnaeus (1792).

> near to or upon the borderlands being equivocal, and endowed with characters which might equally well be assigned to either of the neighbouring species.

Earlier in the letter Leibniz gives the order in the *scala* (from top to bottom): 'Thus men are linked with the animals, these with plants and these with fossils, which in turn merge with those bodies which our senses and our imagination represent to us as absolutely inanimate.' Later, he anticipates the discovery 'of zoophytes, or plant–animals . . . creatures which in some of their properties, such as nutrition or reproduction, might pass equally well for animals or for plants . . .'

Leibniz died early in the 18th century, but as Lovejoy (1936, chapter VI) says 'Next to the word "Nature", "the great chain of being" was the sacred phrase of the 18th century, playing a part somewhat analogous to that of the blessed word "evolution" in the late 19th.' The naturalist most famous for his advocacy of the *scala* (as an *Échelle des Êqtres*) was the Swiss Charles Bonnet, particularly in his *Contemplation de la Nature* (1764). Ritterbush (1964) has turned Bonnet's written account into a diagram (Fig. 3).

There are two components in the concept of the *scala* attributed (of course!) to Plato and Aristotle, respectively. Firstly, the idea of 'plenitude' or, in other words, that creation is a 'plenum': the created world is 'the best of all possible worlds' (to quote Leibniz) and in such a creation there are no gaps between entities — any apparent gaps in the *scala* are simply due to our ignorance. The second component is the *scala* itself. Given the gaps, however, there is a justification for producing an hierarchical classification, using the lacunae to delimit the boundaries between groups; but unless such an artificial classification is declared natural by the tenets of essentialism, there can be no natural concept of homology except as a gradual change in any given structure as one progresses up the *scala*.

The *scala* was regarded as a static phenomenon by 18th century philosophers (including scientists), but the introduction of a worked-out theory of evolution did not dispose of it. Lamarck first introduced his evolutionary theory in lectures in 1800, but his original 'phylogeny' consisted of an '*escalator naturae*', to coin a phrase. Throughout the history of the habitable earth 'monas' was spontaneously generated to evolve through the generations into humans. He admitted the necessity of separate *scalae* for animals and plants and in the immediately following years introduced his theory of 'the inheritance of acquired characters' to explain both deviation from the preordained *scala* and the phenomenon of adaptation. What he did not do, even in the principal statement of his theory, the *Philosophie zoologique* (1809), was to abandon the distinction between elucidating the *scala* ('*distribution general*', a phrase also used by Bonnet), and '*classification*', i.e. using gaps in our knowledge to produce an hierarchical classification. Chapter I of the *Philosophie zoologique* is entitled 'On artificial devices in dealing with the productions of nature — how schematic classifications, classes, orders, families,

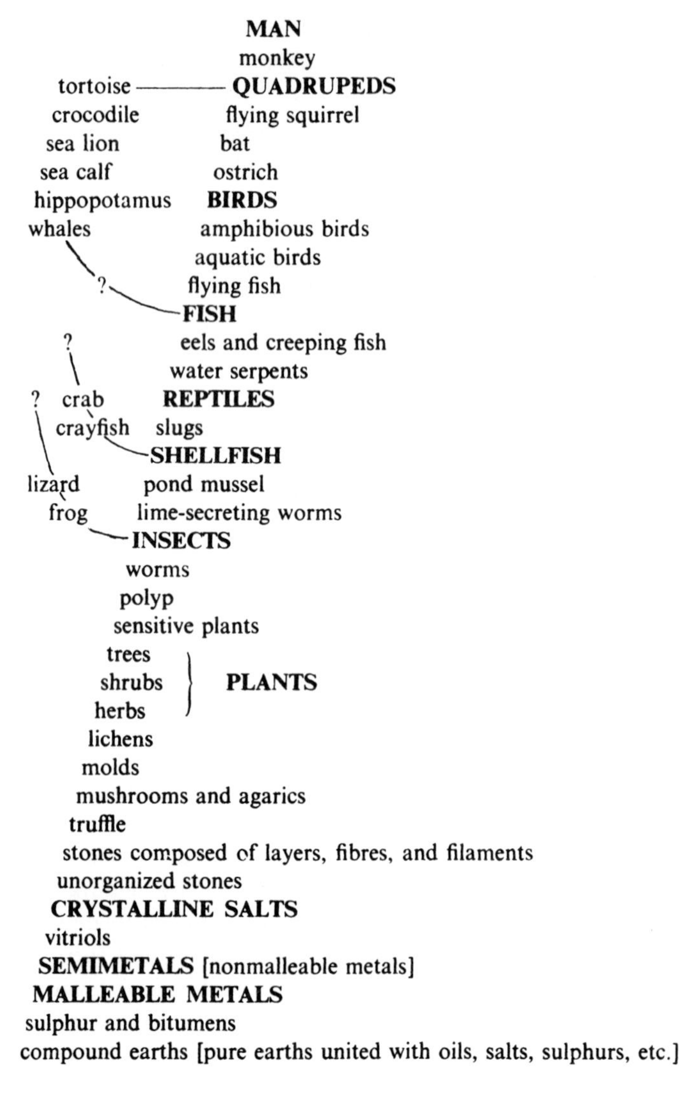

FIG. 3. *Idée d'une echelle des êtres naturelles*. Diagram from Ritterbush (1964), by permission of the publishers. After the written list of Bonnet (1764).

genera, are only artificial devices.' Lamarck did, however, talk of 'affinities', but rather representing the distance between two species on the *scala* (comparable to the 'taxonomic distance' of modern phenetic classification), than representing homology as it did to later 19th century naturalists such as MacLeay (1821) and Strickland (1846) (see Panchen 1994).

The *scala naturae* is with us yet! Much talk of 'lower' and 'higher' animals is a symptom of it, as in the palaeontological conceit of an 'age of fish', followed by an 'age of amphibians', 'age of reptiles' and 'age of mammals'. Darwin used to remind himself not to talk of 'higher' and 'lower' animals, and we find Alfred Russel Wallace (1870, chapter IV) debating whether the Papilionidae (swallowtails) are the highest among butterfly families.

Hierarchy and homology

The most influential early break with the concept of the *scala naturae* was that of Georges Cuvier (1769–1832). In 1800, in *Leçons d'anatomie comparée*, Cuvier simply divided all animals into nine classes, four vertebrate and five invertebrate, but by 17 years later, Cuvier (1817), in *Le règne animal*..., he had settled on his definitive classification of the animal kingdom into four '*embranchements*': vertebrates, molluscs, articulates (arthropods and annelids) and radiates (echinoderms, polyps and various other invertebrates). The basis of this classification is Cuvier's principle of subordination of characters, a direct application of his principle of functional correlation (Russell 1916). The dominant characters are those which both have the greatest constancy, but are also of the greatest importance in determining the functioning of the individual animal. The nature of the nervous system defines the four *embranchements*: a single dorsal trunk with an anterior brain in vertebrates, as scattered masses in molluscs, paired ventral nerve cords with segmental ganglia in articulates, no apparent system (to Cuvier) in radiates.

Given the nature of the nervous system (and the circulatory system that correlates well with it) other vital characters—digestion, locomotor system, etc.—are to some extent predetermined to produce a viable organism. According to Cuvier it is idle to seek any what we would call homologies between different *embranchements*. In this view he was disagreeing with his close colleague and erstwhile friend Geoffroy Saint-Hilaire, a disagreement that led eventually to an acrimonious public debate (Appel 1987). Two other apparently contradictory statements arise from Cuvier's taxonomy: (1) that he made no distinction between homology and analogy; and (2) that he enunciated the principle of taxic homology.

To Cuvier, every feature of the anatomy of an animal was to be interpreted in terms of the viable functioning of the whole; every anatomical difference between two organisms was to be given an adaptive explanation. Unity of plan meant identity of function. But despite this completely teleological approach Rieppel (1988) has argued that to Cuvier homologous characters were those that define taxa. In the terms of modern cladists, taxic homology is synapomorphy (Patterson 1982).

Despite being a colleague in Paris of both Lamarck and Cuvier, Geoffroy Saint-Hilaire worked more in the tradition of the German *Naturphilosophen*, such as Schelling, Meckel, Oken and Carus. The concept of the archetype, later developed for the vertebrate skeleton (Fig. 4) in a spirit of pure Platonic idealism by Richard Owen (1848, 1849), arose in this group (Rupke 1993). But their inspiration appears to have come from Goethe. Interestingly, Goethe's principal preoccupation was not with comparison of corresponding organs or structures in two different species ('special homology': Owen), or comparison between a particular animal and the archetype ('general homology': Owen), but with 'serial homology', despite Goethe's concept of the archetypal plant (*Urpflanze*) and animal (*Urtier*). It was Goethe who suggested that all the appendages of a plant (leaves, sepals, petals, stamens, cotyledons) were variants on a common theme. Similarly, not only were the vertebrae of a chordate serial homologues, but the vertebrate skull was formed of a series of modified vertebrae.

Serial homology was quite compatible with the *scala naturae* as was transformational homology, whether attributed to evolution or not, but taxic homology is predicated on a divergent pattern. Karl Ernst von Baer, despite his background in *Naturphilosophie*, adopted, apparently independently, a pattern of classification similar to Cuvier's, and claimed that even within one *embranchement*, the vertebrates, there was a divergent pattern in ontogeny (von Baer 1828). Early stages in different classes looked similar, later ones less so. There was no *scala* of development. Incidentally, von Baer's demonstration implied another possible explanation for the phenomenon of homology, apart from teleology, essentialism and Platonic idealism. Organs in different animals were homologous because of gradual divergence in development from a common plan (although to von Baer, only within one *embranchement*).

But post-*Origin*, the usual explanation of homology was community of descent as a result of evolution. Ray Lankester (1870) decided that the change needed an overhaul of nomenclature. Owen's (1843) pre-evolutionary definitions, first published in the glossary to his published invertebrate lectures, were:

(1) Analogue: a part or organ in one animal which has the same function as another part or organ in a different animal.
(2) Homologue: the same organ in different animals under every variety of form and function.

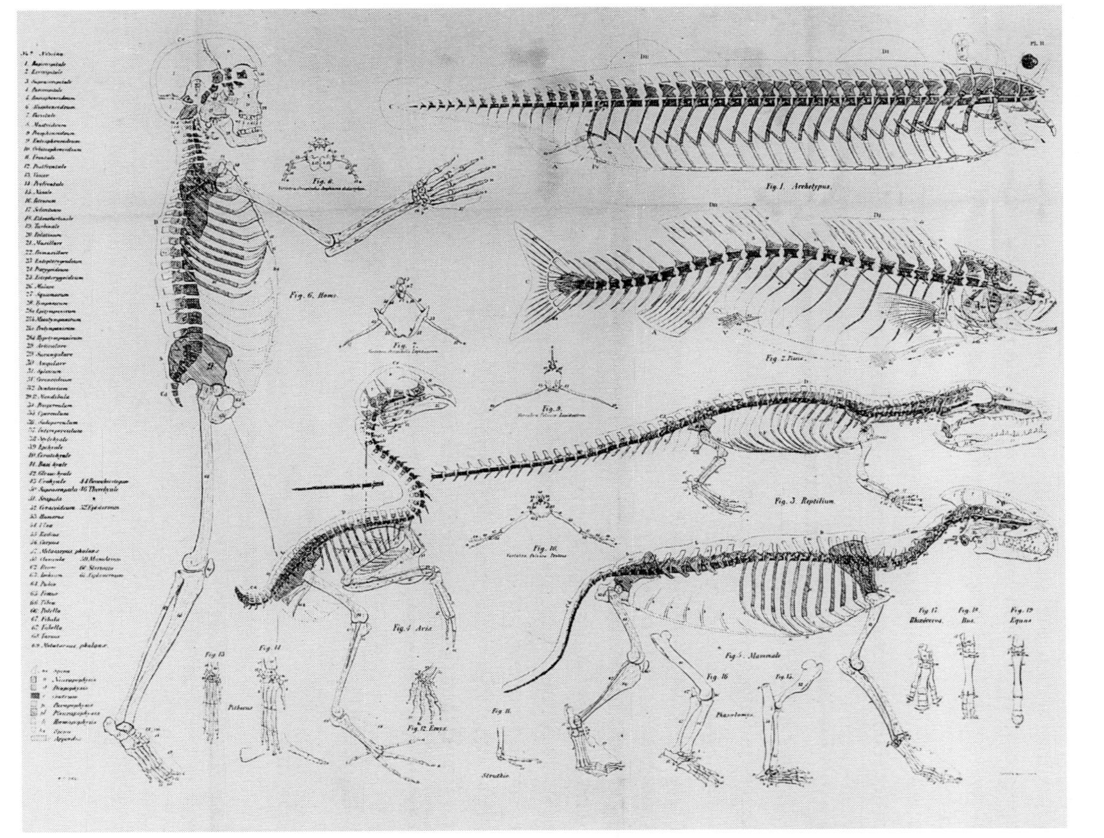

FIG. 4. Richard Owen's archetype of the vertebrate skeleton and its modification in the various classes. After Owen (1848) from the original.

Owen (1848, p.7) was emphatic that the two were not mutually exclusive, indeed to Cuvier they would have been indistinguishable. Lankester's objection was that the term 'homologous' smacked too much of Platonism. 'Special homology' (*sensu* Owen) was to be replaced by 'homogeny':

> Structures which are genetically related, in so far as they have a single representative in a common ancestor, may be called *homogenous*. We may trace an *homogeny* between them, and speak of one as the *homogen* of the other (his italics).

All other resemblances—'serial homology' within a single organism, or analogy between organisms attributed to parallel or convergent evolution, or simple unexplained coincidence—Lankester labelled *homoplasy*: they were *homoplastic*. 'Homogeny' did not catch on, but 'homoplasy' has earned its revival to describe characters incongruent with the pattern of homologies, whether this pattern is interpreted as phylogeny or not.

Since Lankester's time, the concept of homology has been firmly linked to that of phylogeny, except to some 'pattern cladists'. But until recently there has been another (partly hidden) assumption. A hypothesis of homology was assumed to be based on a comparison between structures in (usually) adult organisms. Indeed de Beer (1971) produced a little '*Oxford reader*' tacitly making that assumption and thus pointing out the paradox that obviously homologous structures may have different genotypes, different developmental pathways, different segmental positions and may occur in two different taxa without being ubiquitous in either (Panchen 1994). Now with our greater understanding of molecular biology, we can homologize genes, yet discover that homologous genes need not lead to homologous adult structure (Abouheif et al 1997).

Acknowledgements

I thank Wallace Arthur in particular for his critical reading of the manuscript, and for facilities at the Ecology Centre, University of Sunderland. The Natural History Society of Northumbria allowed photography of their first edition of Belon (1555), and the Robinson Library, University of Newcastle upon Tyne, of Owen's (1848) archetype. The photographs were taken by the Audiovisual Centre, Newcastle University. Figure 2 was photographed with the kind permission of the Linnean Society. The manuscript was typed at Sunderland by Carolyn Stout.

References

Abouheif E, Akam M, Dickinson WJ et al 1997 Homology and developmental genes. Trends Genet 13:432–433

Appel TA 1987 The Cuvier–Geoffroy debate: french biology in the decades before Darwin. Oxford University Press, New York

Aristotle 1945 Parts of animals (trans. AL Peck). Heinemann, London
Belon P 1551 L'histoire naturelle de estranges poissons marins. Regnaud Chaudiere, Paris
Belon P 1555 L'histoire de la nature des oyseaux. Guillaume Cavellet, Paris
Bonnet C 1764 Contemplation de la Nature, 2 vols, 1st edn. Marc-Michel Rey, Amsterdam
Cain AJ 1958 Logic and memory in Linnaeus's system of taxonomy. Proc Linn Soc Lond 169:144–163
Cole FJ 1944 A history of comparative anatomy from Aristotle to the eighteenth century. Macmillan, London
Cuvier G 1800 Leçons d'anatomie comparee de G Cuvier; recueillies et publiées sous ses yeux par C Duméril, vol I. Bandouin, Paris
Cuvier G 1817 Le règne animal distribué d'après son organisation. Deterville, Paris
Darwin C 1859 On the origin of species by means of natural selection, or the preservation of favoured races in the struggle for life. John Murray, London
de Beer GR 1971 Homology: an unsolved problem. Oxford University Press, Oxford (Oxford Biol Readers 11)
Geoffroy Saint-Hilaire E 1818 Philosophie anatomique (vol 1) des organes respiratoires sous le rapport de la determination et de l'indentité de leurs pièces osseusses. JB Baillière, Paris
Lamarck JBPA de M de 1809 Philosophie zoologique, on exposition des considerations relatives à l'histoire naturelle des animaux. Dentu, Paris (English transl: 1914 Zoological philosophy: an exposition with regard to the natural history of animals. J Elliot. Macmillan, London)
Lankester ER 1870 On the use of the term homology in modern zoology, and the distinction between homogenetic and homoplastic agreements. Ann Mag Nat Hist (series 4) 6:34–43
Linnaeus C 1751 Philosophia botanica. Kiesewetter, Stockholm
Linnaeus C 1753 Species plantarum, exhibentes plantas rites cognitas. Laurentii Salvii, Stockholm
Linnaeus C 1758 Systema naturae per regna tria Naturae, tomus 1, 10th edn. Laurentii Salvii, Stockholm (facsimile Brit Mus NH 1956)
Linnaeus C 1792 Praelectiones in ordines naturales plantarum (edited posthumously by Giseke PD). BG Hoffmann, Hamburg
Lovejoy AO 1936 The great chain of being. Harvard University Press, Cambridge, MA
MacLeay WS 1821 Horae entomologicae: or essays on the annulose animals, vol. 1 part II. S Bagster, London
Melville H 1851 Moby-Dick, or the whale. Harper & Brothers, New York
Oldroyd DR 1980 Darwinian impacts: an introduction to the Darwinian revolution. Open University Press, Milton Keynes, p 22–23
Oldroyd DR 1986 The arch of knowledge: an introductory study of the history of the philosophy and methodology of science. Methuen, New York
Owen R 1843 Lectures on the comparative anatomy and physiology of the invertebrate animals. Longman Brown Green & Longmans, London
Owen R 1848 On the archetype and homologies of the vertebrate skeleton. John van Voorst, London
Owen R 1849 On the nature of limbs. John van Voorst, London
Panchen AL 1992 Classification, evolution and the nature of biology. Cambridge University Press, Cambridge
Panchen AL 1994 Richard Owen and the concept of homology. In: Hall BK (ed) Homology: the hierarchical basis of comparative biology. Academic Press, San Diego, CA, p 21–62
Patterson C 1982 Morphological characters and homology. In: Joysey KA, Friday AE (eds) Problems in phylogenetic reconstruction. Academic Press, London (Systematics Association Special Volume series 21) p 21–74
Ride WLD, Sabrosky CW, Bernardi G, Melville RV (eds) 1985 International code of zoological nomenclature, 3rd edn. University of California Press, Berkeley, CA

Rieppel OC 1988 Fundamentals of comparative biology. Birkhäuser, Basel
Ritterbush PC 1964 Overtures to biology: the speculations of eighteenth century naturalists. Yale University Press, New Haven, CT
Rupke NA 1993 Richard Owen's vertebrate archetype. Isis 84:231–251
Russell ES 1916 Form and function: a contribution to the history of animal morphology. John Murray, London
Strickland HE 1846 On the structural relations of organised beings. Philos Mag J Sci (Ser 3) 28:354–364
von Baer KE 1828 Über Entwickelungsgeschichte der Thiere: Beobachtung und Reflexion. Bornträger, Königsberg
Wallace AR 1855 On the law which has regulated the introduction of new species. Ann Mag Nat Hist (new Ser) 16:184–196
Wallace AR 1870 Contributions to the theory of natural selection. Macmillan, London

DISCUSSION

Nicholas Holland: Alec Panchen talked about '*concepts* of homology', which bothered me a little because the term 'concept' is a circumlocution that can be traced back to Adolf Remane. His son, Jürgen Remane, a palaeontologist, wrote an article called '*The development of the homology concept since Adolf Remane*' (Remane 1989). The elder Remane was influenced to a large extent by Immanuel Kant, and for them both homology is something that is understood intuitively, i.e. in the realm of the intelligible world (things as they are) rather than the sensory world (things that are revealed by our senses). This line of thought suggests that it is inappropriate to *define* homology, because we understand it intuitively to begin with. It is interesting that the followers of Remane, such as Rupert Riedl, have defined homology, but in deference to their master they avoided using the word 'definition' and continued to use the word 'concept' instead. Personally, I find it more straightforward to stick to the term 'definition' and not to use the term 'concept', but I would like to know what others think of this. Is there a difference between 'concept' and 'definition'?

Panchen: In my mind there is. I deliberately use the word 'concept' because it is suitably vague. It is important that we should try to find a definition, but then we have to be careful to differentiate among definitions, explanations of the phenomenon of homology and the criteria by which homology is judged. Often, two or more of these are conflated with one another. I hope someone braver than me will provide us with a satisfactory definition.

Wagner: The reason why Riedl refused to define this is because he recognizes that a premature definition will constrain our ability to learn about reality. This can be compared with other scientific concepts. For example, the word 'gene' does not have a definition, but the concept and theory of genetics is extremely rich and powerful. It will not be advantageous for us to come up with a definition of homology now because it is too early and we do not understand the phenomenon itself.

Panchen: What we have now is a succession of unsatisfactory definitions and it seems as though we shall stay in this state forever.

Wagner: If the field turns out to be as productive as the field of genetics, then we won't need a definition.

Rudolf Raff: We do have a historical definition of homology, but a more powerful event happened, i.e. natural selection, so that the governing principle of the last century became the explanation of what creates homology.

Wagner: I would argue that this actively excludes many phenomena that are interesting and related to the concept.

Rudolf Raff: Nevertheless, the focus over the past century has been on the explanation without much thought about the definition. The old definition of Owen (1843) is fine for most of what people have been wanting to explain.

Müller: I agree with Günter Wagner, but I would also like to point out that there is a danger in using the word 'concept' in connection with homology because it seems to indicate that we are merely dealing with a theoretical principle or a taxonomical tool. It conceals the fact that homology is a biological reality. It is a feature of morphological evolution that was used for the recognition of phylogenetic relationships and that led to the establishment of a system of organismal ordering. This system was largely confirmed by later molecular phylogenies. Thus, homology is not merely a concept but a fact of morphological evolution, a fact we still need to explain.

Hall: As Rudolf Raff said, the advent of the theory of natural selection created a major problem in the field of homology because it became necessary to think about it in terms of evolutionary relationships and not just shared structures. This immediately raised the problem of defining the term 'concept'. I'm not sure whether most of us would follow Mayr's definition of 'the same structure in animals that share a common recent ancestor'. What does 'common, recent ancestor mean'? How common is common, and how recent is recent?

Maynard Smith: I don't want to suggest to you what the definition of homology ought to be, but I would say that you are spitting in the wind if you imagine that you can define it as you like. I didn't know what homology was, so yesterday I looked it up in the Collins *Dictionary of biology* just to see what we were going to discuss. It doesn't mention Owen's concepts at all, only 'common descent'. I'm not saying that this is the correct definition, but it is the meaning that people use. If you're going to use the word to mean something else, people will not understand what you're saying. This is not only true for homology, it is true of many scientific terms, including 'gene'. All one can do is say 'in this particular context what I mean by homology is . . . ' or not use the word at all, but you can't decide on a definition and insist that everyone should use it, because they won't.

Akam: But the problem has turned full circle. For the last 100 years people have been interested in working out patterns of relationship and recognizing homology due to common descent. However, there is now a focus on homology as a reflection of common internal principles of organization, which we see in terms of genes but which Owen saw in terms of morphology. Therefore, we should look back at the historical meanings and usages of the term, and see how those relate to the questions that we grapple with in studies of common regulatory genes and common developmental pathways.

Wagner: I agree. One of the reasons why we are finding it difficult to reach a conclusion is because there is an inherent tension between the phylogenetic definition of homology and the meaning of homology as character identity. It is rarely recognized that the phylogenetic definition of homology presupposes that we know what 'sameness' means, and we don't. It is now becoming interesting because we are studying the developmental and genetic bases of character identity and why certain organizational principles are stable, natural units of biological organization. If we don't keep these two aspects separate we will remain confused.

Panchen: This is the point I was trying to make about the difference between definition and explanation. We have a satisfactory explanation but we have no agreed definition of sameness.

Wagner: I have argued elsewhere (Wagner 1994) that we have an explanation for the taxonomic distribution of homologues by inheritance from common ancestors: it's not an explanation of homology *per se*, it refers to the taxonomic distribution of homologues.

Akam: One can envisage situations where an organizational homology is generated that is not dependent on any sort of lineal ancestor; for example, the transfer of a developmental process or network to a new structure, thus producing an internal organizational homology.

Wagner: This has been known for many years, but it was pushed aside because of the dominance of phylogenetic thinking.

Wilkins: It seems to me that you would like to collapse the distinction between homology and homoplasy. In several of your recent articles you suggest that the distinction is insignificant or even false, whereas many people feel that it is important. We need to know whether things are similar because they have a common ancestry or because they happen to use similar genetic machinery even though they came from different ancestors.

Wagner: We have to ask what it is important for. Scientific concepts are only as good as the purpose for which they are designed. Therefore, if we are clear about what we want to do with the concept, then these problems disappear. In phylogenetic research it is important to distinguish between genealogical homology and homoplasy, but it doesn't cover all the other biological problems that exist concerning character identity, repeated elements, cryptic homology, etc. It is an important distinction, but I would warn against collapsing all the biology into phylogeny, as cladists sometimes do. We are addressing different types of problems and trying to explain different aspects of biological reality. We must recognize that there is a problem with character identity from an organizational and developmental point of view.

Maynard Smith: I don't want to shove this problem under the carpet. It is a valid problem, but all I'm saying is that if you use the word homology, you don't know how it will be understood. I invented a term some years ago, and I have absolutely no control over the way it is used.

Akam: In the field of molecular biology, homology is often used to mean similarity without the implication of common evolutionary origin, but if we simply accept this common usage we fall into a mire of ambiguity.

Rudolf Raff: One of the problems with what Günter Wagner is saying is that we are now beginning to address issues that involve criteria rather than issues that involve definition. The latter have to be tied to phylogenetic questions of origin. Problems will arise if non-linear transformations are involved, i.e. transformations of things that are homologous in some evolutionary sense but are no longer recognizable.

Wagner: Absolutely. Whenever the field of comparative biology makes a major step forward, the concept of homology is replaced by terms that fit more precisely the needs of the field. For instance, with the advance of molecular genetics it became clear that we needed to distinguish different forms of homology between genes, such as orthology and paralogy (see also Holland 1999, this volume). As we acquire a deeper understanding of the mechanistic underpinnings of character identity we will have to introduce additional distinctions. For instance, there is character identity in the narrow sense, like the homology between the limbs of different primate species. They only differ in minor details and are in the true sense 'the same organ in different species' as Owen said. On the other hand, there is the relationship between characters that are not 'the same' but where one character is phylogenetically derived from the other. For example, the forebrain cortex (iso-cortex) of mammals is phylogenetically derived from the so-called dorsal cortex of 'reptiles' (Striedter 1997) even if the iso-cortex is not the same as the iso-cortex in its structure and organization. Both relationships, character identity and phylogenetic derivation have been called homology, even if they refer to different biological situations. If we do not distinguish between character identity in the narrow sense and the transformation of one character into a different one, we will not make sense.

Hall: The mammalian ear ossicles that arose from elements of the lower jaws of reptiles is an good example of where the element of sameness has disappeared.

Wagner: But there is still a derivational relationship between the ear ossicles of mammals and the jaw joint of reptiles, and this relationship has been called homology for good reasons. But these two characters are not the same in any biological sense.

Rudolf Raff: But those are still issues of criteria for recognizing homologues, and they are not part of the definition of homology. We are trying to understand why there is a historical, informational connection between structures of the past and present. It is the basic idea that somehow that there's an underlying truth to homology regardless of all these other conflating issues.

Lacalli: I would like to elaborate on what Günter Wagner said. The homology concept should be considered as a tool, and how that tool is used in a given instance is dependent on the particular data set being examined. For example, a number of the talks at this symposium concern the vertebrate skeleton. Here one has detailed morphological data from a vast array of both fossil and living forms, so the data set is rich, and it's not surprising that precise definitions and a subsidiary terminology are needed. In the case of molecular phylogenies, there is a further problem of the misuse (or too broad use) of the term 'homology', so we again need to be careful about terminology. In contrast, among the basal phyla of the animal kingdom, where there are huge gaps in the phylogeny and no prospect of a fossil record, we have few data and all sorts of unresolved controversies over which, if any, characters are truly primitive. In such cases we have no choice but to use the homology concept in a broad way. The value of the term 'homology' in practice seems to me, therefore, to be its flexibility. A precise definition is not needed, because new terms can be invented for specific data sets as required. But we also have to exercise a degree of tolerance, as scientists, so as not to force our own preferred set of terms and methodologies on others, who may be dealing with data of a fundamentally different kind.

Abouheif: I disagree with the notion that the definition of homology should be tailored to specific data sets. In the majority of cases we should be able to use the term homology and define it in such a way so that it can be used operationally, otherwise the term becomes somewhat metaphysical.

Tautz: But it is not the definition that is important, it is the criteria of how to establish the homology of given characters. Remane has established criteria for identifying homologous characters. These criteria include: (1) identity of position; (2) identity of specific morphological quality; and (3) bridging through intermediate forms, which may be ontogenetic stages or similar structures in related species (Remane 1952). But he has refrained from giving a definition of homology.

Roth: Irrespective of whether we want to generate new terms or not, using the concept of 'homology' as a tool is important if we want to communicate relationships. We may not want to argue over the use of a word, but rather we should be more concerned about communicating which processes occur in which relationships.

Striedter: But confusion arises over the use of the term 'criteria of homology'. In my opinion, there is only one testable criterion of homology, and that is phylogenetic continuity. What people typically talk about when they discuss criteria of homology are Remane's criteria of topology, connections, etc., and to me those are criteria of character identity. Some of the characters identified as 'the same' on the basis of Remane's criteria may be homologous in the sense that there is phylogenetic continuity between them, but others may not be homologous in this sense—they may have arisen as a result of homoplasy.

Tautz: But his third criterion, the bridging through intermediate forms, has a phylogenetic component.

References

Holland PWH 1999 The effect of gene duplication on homology. In: Homology. Wiley, Chichester (Novartis Found Symp 222) p 226–242
Owen R 1843 Lectures on the comparative anatomy and physiology of the invertebrate animals. Longman Brown Green & Longmans, London
Remane A 1952 Die Grundlagen des natürlichen Systems, der vergleichenden Anatomie und der Phylogenetik. Akademische Verlagsgesellschaft, Geest & Portig K-G, Leipzig
Remane J 1989 Die Entwicklung des Homolgie-Begriffs seit Adolf Remane. Zool Beitr 32:497–503
Striedter GF 1997 The telencephalon of tetrapods in evolution. Brain Behav Evol 49:179–213
Wagner GP 1994 Homology and the mechanisms of development. In: Hall BK (ed) Homology: the hierarchical basis of comparative biology. Academic Press, San Diego, CA, p 274–299

Homoplasy, homology and the problem of 'sameness' in biology

David B. Wake[1]

Museum of Vertebrate Zoology, University of California, Berkeley, CA 94720-3160, USA

Abstract. The reality of evolution requires some concept of 'sameness'. That which evolves changes its state to some degree, however minute or grand, although parts remain 'the same'. Yet homology, our word for sameness, while universal in the sense of being necessarily true, can only ever be partial with respect to features that change. Determining what is equivalent to what among taxa, and from what something has evolved, remain real problems, but the word homology is not helpful in these problematic contexts. Hennig saw this clearly when he coined new terms with technical meanings for phylogenetic studies. Analysis in phylogenetic systematics remains contentious and relatively subjective, especially as new information accumulates or as one changes one's mind about characters. This pragmatic decision making should not be called homology assessment. Homology as a concept anticipated evolution. Homology dates to pre-evolutionary times and represents late 18th and early 19th century idealism. Our attempts to recycle words in science leads to difficulty, and we should eschew giving precise modern definitions to terms that originally arose in entirely different contexts. Rather than continue to refine our homology concept we should focus on issues that have high relevance to modern evolutionary biology, in particular homoplasy—derived similarity—whose biological bases require elucidation.

1999 Homology. Wiley, Chichester (Novartis Foundation Symposium 222) p 24–46

Why are we still talking about homology?

I will grant that someone might be able to generate an original thought concerning homology, but I doubt it. When Hall (1994) appeared, I thought we were at the end of the line. Homology is important. Indeed, I argued that the general concept of 'sameness' was fundamental to the entire modern science of biology, which relies so heavily on model organisms (Wake 1994). Now we know about homology; we understand its importance as a general, even baseline, concept. Isn't it time to move on?

[1]Due to illness, David B. Wake was unable to attend the symposium, and Adam Wilkins presented this chapter in his stead.

There are several reasons for my stance. First, homology is seen by some to constitute the most fundamental kind of evidence of the existence of evolution, and this has led critics to raise challenges to specific instances of homology and even to challenge the basic idea on the grounds that biologists cannot agree on a definition of homology. If practitioners can fight about homology after all these years, the argument goes, perhaps there is nothing to it after all. Second, I detect a kind of modern-day essentialism creeping into the debates about homology. I, too, once sought the 'biological basis of homology' (Wagner 1989), and I remain fascinated by the general phenomenon of stasis in evolution (Wake et al 1983). There are genuine biological questions involved in these issues. But the risk is increasing that we will make homology something that it is not, as for example when we explore ideas such as the latent homology of organs (insect wings and tetrapod limbs) that might share elements of a developmental programme but are phylogenetically independent in origin (Shubin et al 1997). Finally, specialized fields of biology, such as phylogenetics, have seized the concept of homology for their own, even though the wise founder of cladistic methodology (Hennig 1966) saw well the intellectual quagmire in front of him and opted for a technical approach and specialized terms.

I wonder if we have made any progress in dealing with homology since the late 19th century, when the German morphologists accepted Darwinian principles. Gegenbaur (1870) recognized that the proper discrimination of homologies from analogies required the study of genealogical relationships. Berg (1926), in contrast, believed that genealogical matters could only be resolved on the basis of character data and that character analysis was necessarily prior to establishing hypotheses of relationship. Thus, we see the two main perspectives on homology that are with us today. On the one hand, systematists, and in particular cladists, focus on common ancestry and taxa (the systematic concept of homology, Sluys 1996), while morphologists focus on similarity and the developmental individuality of putative homologous structures (the biological concept of homology, Sluys 1996, Wagner 1989). This is close to a true dichotomy, with one group viewing homology of any given structure to be strictly dependent on the outcome of a formal cladistic analysis (i.e. there is a test of homology), while the other group views its task as the explanation of the maintenance of structural identity during the course of evolution. Often the terms taxic and transformational homology distinguish the two approaches (Rieppel 1988); one group focuses mainly on pattern while the other is concerned with process. While some authors (e.g. Sluys 1996) envision a complementarity in these two approaches, with one group (the biological homology approach) emphasizing the causes of what the other (the systematic homology approach) views as symplesiomorphies and homoplasies—features irrelevant to its search for a robust phylogeny which would reveal the synapomorphies, the true homologues—I have a different view. I think it is time we focus on scientific research programmes and specific questions and avoid the fruitless quest for a compromise, or an accommodation, when none is likely to be forthcoming that will be stable through time.

Why the homology debate is a distraction

From time to time I have engaged in discussions with individuals who question aspects of evolutionary theory. These are usually professionals in other disciplines, but occasionally they are biologists who are not convinced that the modern synthetic theory of evolution is well established. It is interesting to determine what these often intelligent and even well-informed individuals see as the weak points in evolutionary theory. Often ancient words and concepts are the problem, such as adaptation, species and, of course, homology. These are pre-Darwinian concepts that we have recycled and massaged, attempting to save vague old concepts by clothing them in modern perspective. A molecular biological colleague once asked me, in frustration: 'how can we take you evolutionists seriously when you can't even agree on what a species is?' Evolutionary theory demands that there be entities, species, and the fact that there has been both branching and extinction means that the biological continuum is broken up into fragments of the genealogical nexus, and we continue to find it convenient to term these species. It is a separate, specialized task to determine the bounds of species. Adaptation was a term that served reasonably well as a general concept until we started trying to be precise and exact (e.g. Williams 1966). Then we came to understand that we wanted the term to be multifunctional (e.g. adaptation as a state of being, adaptation as a process, adaptation as an outcome). We tried to reify the concept by introducing new terms (e.g. exaptation, Gould & Vrba 1982), and finally, as happened with homology, it became clear that adaptation was only understandable in any exact sense in a phylogenetic context (e.g. Greene 1986). But scientific progress is related mainly to the development of specific research programmes not targeting adaptation but rather processes and patterns of evolution in general (many examples in Rose & Lauder 1996). When doubters, including creationists, raise questions about adaptation they are concerned with origins of novelties, not with the inner workings of evolutionary processes and the patterns that result.

Homology is subject to these same problems. We have taken an ancient term, accepted it as real, and then reified it to serve our present purposes. Doubters point out that if homology is so important, we all should agree as to what it is. One main criticism is that, like species and adaptation, there is no 'naturalistic' mechanism to determine what species, adaptations or homologues really are. Where is the empirically demonstrated naturalistic mechanism of homology? This is exactly the kind of question that we cannot answer, but more to the point, it is an irrelevant question. The clear implication is that if there is no answer evolutionary theory is in trouble. But the central issue is overlooked—homology is the anticipated and expected consequence of evolution. Homology is not evidence of evolution nor is it necessary to understand homology in order to accept or understand evolution. That there is a genealogy of life on this planet necessitates that there be homology. Every component of an organism is involved, from nucleic acid base pairs to behaviours and even interactions between two or more genealogical entities (such as between host and parasite). Two organisms have the same organ because their common ancestor had the organ. It is a genealogical necessity, and no naturalistic mechanism is necessary to account for the phenomenon other than inheritance. Darwin knew this, and Gegenbaur accepted it, but we are still stuck somewhere between Darwin and Owen, knowing that homology is about evolution, but still wanting a naturalistic explanation. The only way out of this dilemma is to stop talking about homology and instead deal with the real questions that interest us. Systematists should adopt the terminology of Hennig (1966), who recognized that exact science requires precise, technical terminology. So-called homology assessment, however, is an essentially *post hoc* process in systematics, and there is no need to use the word homology. Besides, who wants homologues that can become homoplasies simply with the addition of another character to a matrix, or even in an alternative, equally parsimonious tree? Those interested in so-called biological homology should shift their attention to what I consider to be the real questions relating to similarity in form during the evolution of life.

Some futile exercises

Is homology absolute? That is, can homology be seen as being independent of a theory of phylogenetic relationships? No! As de Pinna (1996) has appropriately observed, absolute homology lacks any kind of theoretical foundation.

Can homologues be partial? Of course they can (Hillis 1994)! Consider the *Hox* gene system as a vivid example. In fact, one can argue that all homology among genealogical entities is likely to be partial at some level, except between the closest relatives. Homology is absolute only when we make it so methodologically, as when we identify characters and states, assume their homology and even 'verify' it with phylogenetic analysis. We forget that we have created the intellectual framework.

Are iterative structures homologues? Of course not! This is a 'levels of analysis' problem, a hierarchical issue (Lauder 1994). What causes iteration might well be homologous in different taxa, but the iterated parts are simply that.

Can homology ever be definitively demonstrated? I think not. I view homology as something that follows from the fact of evolution. There is a continuity, as

Darwin perceived, and branching and extinction have produced genealogical entities whose phylogenetic relationships are inferred largely on perceived degree of overall homology, especially that portion of overall homology that is uniquely shared. Depending on our outlook we may posit that organs are homologous on some biological grounds and be happy when the proposition passes a phylogenetic test (as with the transition from bones forming a lower jaw articulation in some amniotes to forming inner ear ossicles in others), but all that we have done really is identified a pathway of phylogenetic continuity. All proximal attempts to explain homology in terms of structure, connectivity, topography or morphogenesis, for example, can fail. Is homology any more than phylogenetic continuity? Maybe we need to turn Van Valen's (1982) definition ('correspondence caused by continuity of information') around: the continuity of information necessitates correspondence.

Some real questions

Stasis

That which Darwin termed unity of type and what Eldredge & Gould (1972) revived as a modern problem remains unexplained. At one level it is simple enough—inheritance mechanisms assure that descendant taxa continue to resemble ancestors, often in nearly every detail of structure. But natural selection theory tells us that local adaptation should be taking place, and so there must be some mechanistic explanation for the resistance of organisms to evolve (Wake et al 1983). Do any of the many suggested causes offer a general explanation for morphostatic mechanisms (those which maintain structure in evolution, contrasted to morphogenetic mechanisms which generate structure, Wagner 1994)?

Modularity

This leads to what has been termed iterative or serial homology. Some truly exciting research is taking place dealing with this large topic (see Raff 1996 for a general review). Modularity occurs at many levels of biological organization (like most issues relating to homology, hierarchical thinking is necessary), and the proximal (genetic and developmental) foundations for it are slowly being unravelled. The *Hox* gene system and its relation to brainstem organization, the formation and iteration of vertebrae, and limb development in vertebrates is an excellent example of how we are coming to understand one intricate system. We can construct gene trees, follow paralogous and orthologous changes, and come to understand how iteration and evolution have produced a genetic signalling system that appears to be central to vertebrate design. In turn, parts of the system are deployed during morphogenesis to signal and perhaps direct events taking place in diverse black boxes that are beginning to be opened. Modularity occurs in this system at levels from gene sequences to discrete organs such as vertebrae.

How modules form and get organized is still largely unknown. My current favourite is a 5 kb fragment of DNA related to known retrotransposons of fungi that is repeated about 10^6 times in the genome of the plethodontid salamander *Hydromantes* (Marracci et al 1996). What is especially mysterious about this case is that the large genome of this salamander has severe developmental and morphological consequences, leading to extensive homoplasy in the form of simplification of brain tissue-level organization, for example (Roth et al 1997). Modularity at one level of organization frequently has consequences at other levels, and study of such systems will be enlightening for understanding how form evolves.

The similarity of structure, often down to fine details, of the fore- and hindlimbs of tetrapods has been a challenge since the time of Owen. This is a classic case of iterative homology. The parallel evolution of the two limbs can be stunning, and is nearly the same degree of similarity as bilateral symmetry (another morphological topic worthy of much more attention than it has received). For example, in the evolution of the Caudata two distal mesopodial elements (distal carpals 1 and 2 in the forelimb and distal tarsals 1 and 2 in the hindlimb) fused to form the basale commune, a structure universal in living salamanders and found nowhere else (a synapomorphy of Caudata, Shubin et al 1995). There are two views of how this event occurred. Either a caenogenetic (adaptive) event associated with precocial use of developing digits occurred, in which case a fused element provides a more substantive base for the first two digits, which function to some degree as sense organs in larvae (Marks et al 1998), or this is a neomorph, a secondary formation of the first two digits from an unknown ancestor which had undergone digital reduction and had only two or three digits (Wagner et al 1998). Whatever the evolutionary pathway to fusion, the fact remains that the same fusion occurred in both fore- and hindlimb. The adaptive explanation would apply only to the forelimb, so the hindlimb would have co-opted the forelimb developmental change. Alternatively, all living salamanders that have undergone limb or digital reduction follow different pathways in the fore- and hindlimbs (for example, having different numbers of digits on fore- and hindlimbs, or reducing the number of digits on one set but not the other) so it seems likely that if digits were re-evolved they probably arose from different patterns. Why they should be the same is unclear, but this is a repeated story in vertebrate limb evolution.

Progress has been great in understanding gene expression patterns in relation to structure and evolution of the vertebrate limb (e.g. Sordino & Duboule 1996) and the mechanics of limb development (Marks et al 1998). Limb development is a research topic where we have moved beyond questions of homology to approach an understanding of how form has evolved.

Preservation of design

Lineages evolve distinctive designs that persist in the face of genetic and developmental change. One of the first clear examples was highlighted by Berg (1926)—Spemann's discovery that the lens of salamanders, which forms from the ectoderm at the spot where the optic vesicle adjoins the head, when removed and allowed to regenerate forms not from cornea but from the upper margin of the iris, an entirely different kind of tissue. Are the two lenses, in the same organism, homologous? There are many such puzzles in experimental embryology (recall the oft-cited example that the dorsal hollow nerve cord forms by infolding in most taxa but by cavitation in several). In salamanders the notochord plays a critical role in the formation of vertebrae, which will not form in its absence. However, in the regenerating salamander tail everything is present, including fully formed, individuated vertebrae, except the notochord!

Some persistent puzzles relate to the numbers of digits in vertebrates and their identities. This is especially the case in comparing early tetrapods with more than five digits with subsequent tetrapods, which virtually never have more than five (Coates 1993), and in comparing living birds with their presumptive theropod dinosaur ancestors (Burke & Feduccia 1997 and subsequent commentary). What is the mechanistic basis for determination of digital identity? I have studied this problem in salamanders, from two perspectives: first, an unusually large member of a four-toed lineage that produced five toes in an unusual manner; and second, an unusual member of a five-toed species that produced four external toes, but which had two skeletons in the last toe of one foot and one in the last toe of the other (Wake 1991). To attempt to homologize such digits and say which is which would be futile, whereas it would be illuminating to understand the morphogenetic processes responsible for longitudinally splitting a digital primordium. I suspect that basic developmental mechanics is involved and that in these large-celled organisms the numbers of cells available to form a condensation and then available to form a bifurcation is critical. In the four-toed example, it seems clear that the digit present is neither number 4 nor number 5, but both.

Latent homology and homoplasy

The condition of permanent larvae in salamanders has arisen many times independently, and there are species-level taxa that never metamorphose. We do not know if a non-metamorphic species has given rise to a species that metamorphoses but it is certainly conceivable. Some bones never form in larvae, but even should a permanently larval species reproduce, the bones remain as latent elements. Imagine that maxillary, septomaxillary and prefrontal bones, all absent in plethodontid salamander larvae, reappear in a derivative species. These would be identified as homoplasies (in this case, reversals) in phylogenetic analysis, but a morphologist would insist that the tooth-bearing maxillary bone is the same bone as in distantly related salamanders. It must be a homologue! It is in the same place, it has the same form, it develops in the same way and in the same sequence, and it has the same relationship to neighbouring elements. What has been retained is the entire morphogenetic system. In contrast, one can imagine that only one or a few genes were involved in failure to metamorphose, and the time involved in the shutdown of the morphogenetic system may have been too short for the system to have undergone random mutational decay. This is, in short, a levels of analysis problem. The maxillary bone both is and is not a homologue. It does not need to be renamed. This is a simple example of a phenomenon that is extraordinarily important in evolutionary morphology.

Homoplasy is derived similarity that is not the result of common ancestry. The biologist who had an early clear vision of homoplasy and its significance in evolution was Berg (1926), who argued that the same 'laws' of development governed ontogeny and phylogeny. If we replace 'laws' with rules or regularities, and admit a fuller role for natural selection than Berg was willing to do, we approach a modern conception. We are interested in homoplasy because of what the phenomenon can teach us about the evolution of form. The topic has received a recent book-length treatment (Sanderson & Hufford 1996) and I have given my own perspective on the issue (Wake 1991) so I will not belabour the point here. I simply note that while investigating homoplasy we have an opportunity to study the persistence of morphogenetic pathways and ontogenetic trajectories, with related organisms progressing to differing extents along these and thus expressing variations on essentially the same theme. I see roles for functionalist (that is, selectionist or externalist) and structuralist (that is, mechanistic, developmental–genetical or internalist) approaches to the general problem, which always must have historical (phylogenetic) rooting.

Directionality in evolution?

Lineages do appear to have evolutionary directionality. There is a kind of evolutionary canalization that arises from the shared morphogenetic–morphostatic mechanisms that characterize lineages, so that lineages tend to travel in evolutionary pathways of least resistance. This, again, is a topic of great interest to Berg (1926), but one that has received far too little attention in recent years. It is time to return to this important issue.

Conclusions

We have a job before us. As students of form we must not be distracted by sterile debates over matters of interest mainly to students of the history of biology. Rather, we need to keep our eyes on the target—how does form evolve? Why do some structures evolve again and again and again, even iteratively in the same organism? Why do some structures, even entire phenotypes, persist in the face of ever-changing environments and great molecular change? What are the hierarchical interactions in biological systems that produce form and why is form so limited in its expression? Never has there been so much reason for optimism that we will finally come to understand how form evolves. Let us maintain focus and ask answerable questions.

Acknowledgements

I thank Marvalee Wake for discussion of issues raised in this essay and for her comments on the manuscript.

References

Berg L 1926 Nomogenesis or evolution determined by law. Constable & Co, London (English translation of 1922 Russian edition)

Burke AC, Feduccia A 1997 Developmental patterns and the identification of homologies in the avian hand. Science 278:666–668

Coates MI 1993 Ancestors and homology (The origin of the tetrapod limb). Acta Biotheor 41:411–424

de Pinna M 1996 Comparative biology and systematics: some controversies in retrospective. J Comp Biol 1:3–16

Eldredge N, Gould SJ 1972 Punctuated equilibria: an alternative to phyletic gradualism. In: Schopf TJM (ed) Models in paleobiology. Freeman, Cooper & Co, San Francisco p 82–115

Gegenbaur C 1870 Grundzüge der vergleichenden Anatomie, 2 Aufl. Leipzig

Gould SJ, Vrba E 1982 Exaptation—a missing term in the science of form. Paleobiol 8:4–15

Greene HW 1986 Diet and arboreality in the emerald monitor, *Varanus prasinus*, with comments on the study of adaptation. Fieldiana Zool New Ser 31:1–12

Hall BK (ed) 1994 Homology: the hierarchical basis of comparative biology. Academic Press, San Diego, CA

Hennig W 1966 Phylogenetic systematics. University of Illinois Press, Urbana, IL

Hillis DM 1994 Homology in molecular biology. In: Hall BK (ed) Homology: the hierarchical basis of comparative biology. Academic Press, San Diego, CA, p 339–368

Lauder GV 1994 Homology, form, and function. In: Hall BK (ed) Homology: the hierarchical basis of comparative biology. Academic Press, San Diego, CA, p 151–196

Marks S, Wake DB, Shubin N 1998 Limb development and evolution in the salamander genera *Desmognathus* and *Bolitoglossa* (Amphibia, Plethodontidae): separating hypotheses of ancestry, function and life history, submitted

Marracci S, Batistoni R, Pesole G, Citti L, Nardi I 1996 *Gypsy/Ty3*-like elements in the genome of the terrestrial salamander *Hydromantes* (Amphibia, Urodela). J Mol Evol 43:584–593

Raff RA 1996 The shape of life: genes, development and the evolution of animal form. University of Chicago Press, Chicago, IL

Rieppel O 1988 Fundamentals of comparative biology. Birkhäuser Verlag, Basel

Rose M, Lauder GV 1996 Adaptation. Academic Press, San Diego, CA
Roth G, Nishikawa KC, Wake DB 1997 Genome size, secondary simplification, and the evolution of the brain in salamanders. Brain Behav Evol 50:50–59
Sanderson MJ, Hufford L (eds) 1996 Homoplasy: the recurrence of similarity in evolution. Academic Press, San Diego, CA
Shubin N, Wake DB, Crawford AJ 1995 Morphological variation in the limbs of *Taricha granulosa* (Caudata: Salamandridae): evolutionary and phylogenetic implications. Evol 49:874–884
Shubin N, Tabin C, Carroll S 1997 Fossils, genes and the evolution of animal limbs. Nature 388:639–648
Sluys R 1996 The notion of homology in current comparative biology. J Zool Syst Evol Res 34:145–152
Sordino P, Duboule D 1996 A molecular approach to the evolution of vertebrae paired appendages. Trends Ecol Evol 11:114–119
Van Valen LM 1982 Homology and causes. J Morphol 173:305–312
Wagner GP 1989 The biological homology concept. Annu Rev Ecol Syst 20:51–89
Wagner GP 1994 Homology and the mechanisms of development. In: Hall BK (ed) Homology: the hierarchical basis of comparative biology. Academic Press, San Diego, CA, p 273–299
Wagner GP, Kahn PA, Blanco MJ, Misof B 1998 Evolution of *Hoxa-11* expression in amphibians: is the urodele autopodium an innovation? Am Zool, in press
Wake DB 1991 Homoplasy: the result of natural selection, or evidence of design limitations? Am Nat 138:543–567
Wake DB 1994 Comparative terminology. Science 265:268–269
Wake DB, Roth G, Wake MH 1983 On the problem of stasis in organismal evolution. J Theor Biol 101:211–224
Williams, GC 1966 Adaptation and natural selection. Princeton University Press, Princeton, NJ

DISCUSSION

Hall: Did David Wake give any examples of partial homology?

Wilkins: No, he just asserts that it's true. One example, however, may be when genes have different exons from different sources.

Wagner: Another may be that, for example, paired appendages of vertebrates, or at least gnathostomes, share certain developmental mechanisms, such as those which determine anteroposterior patterning, but they differ with respect to skeletogenesis (see Wagner 1999, this volume). Therefore, at the paired appendage level, there is some degree of homology, but less so than the homology between the pectoral fins of two fish species or the limbs of two salamander species, for example. In that sense, there is a hierarchy of shared, derived characters that could be interpreted as levels of homology.

Hall: But David Wake referred to partial homology at a particular level, which is different to the situation you have described.

Wagner: It is still at the level of the structure of the paired appendage, but paired appendages from different groups of species (taxa) have different degrees of shared characters, and this leads to a higher degree of homology among fins than between fins and limbs.

Wilkins: When you say a higher degree of homology, do you mean greater similarity?

Wagner: Yes, but not in the sense of simple resemblance. As I have argued elsewhere (e.g. Wagner 1996), the homology concept captures the fact that most closely related characters share developmental constraints. For instance, the pectoral fin of a lion fish is not similar to the pectoral fin of a zebrafish, but they are still the same character because both evolve under the same set of developmental constraints. If we compare them to the pectoral fins of a lungfish, the number of developmental constraints is smaller. There can be dissimilarities that evolve within a constant framework of developmental constraints (one could call them 'variations on a theme') or more fundamental dissimilarities that lead to different developmental constraints on evolution (one could call them 'different themes'). Partial homology is partial overlap in developmental constraints, not just a lower degree of similarity in appearance.

Wray: Is this so different from simply stating that in two different lineages structures have diverged from their initial similar state?

Wagner: No, but it matters in what respect the characters have diverged. Only if the variational properties of the character have changed (i.e. developmental constraints and developmental degrees of freedom) is it useful to talk about partial homology. It is not sufficient that a character changes its size or colour if these differences are easily reversible and they do not change the constraints for further evolution.

Abouheif: The comparison of fish with tetrapod limbs may be a good example of partial homology. Some of the proximal elements, such as the radius and the ulna, are conserved. However, in the tetrapod limb, although the proximal elements are conserved, the distal elements, such as the digits, are morphological novelties. Therefore, there is partial homology in that some of the elements are homologous, whereas others are not.

Roth: Some would argue that the idea of partial homology is an issue of character delineation, i.e. if you subdivide the character in a precise enough fashion you will find an entire component that is homologous and another that isn't. However, Rolf Sattler (1994) has argued that some characters, especially botanical ones, are so continuously melded together that is difficult to do a one-to-one comparison. Rather, it is necessary to talk in terms of percentages or continuous numerical differences and similarities.

I also have a question. You seemed to suggest that David Wake said that the debates about homology have been fruitless and we should move on to address more useful questions. Yet before moving away from those questions, in his introduction he seems to suggest answers to them himself. Do you take him to be saying 'these are the answers' or simply 'here are some definitions that we can work from'?

Wilkins: I was confused as to what elements of his critique he felt were the most important. On the one hand, he says the idea of homology is so simple that we needn't bother with it because it's a direct consequence of relationship through descent. But on the other hand, he says that it is a muddied concept because it has been used in different ways, and we should therefore leave it behind and not reify it. It seems to me that he gives two reasons for abandoning it or, at least, debates about it: (1) it's too simple to bother with; and (2) it's too confused to bother with.

Müller: I'm sure that it is not his intention to abandon the concept of homology. He probably means that the ongoing discussion of homology is becoming futile, and that we should accept homology as a biological fact.

Wilkins: It is important as a concept because we still need to determine whether certain characters are homologous, but he apparently feels that there has been too much emphasis on it. Instead of spending a lot of time debating whether certain characters are homologous or not, and trying to be specific about what that means, we should concentrate on looking into the processes that generate characters.

Striedter: It is interesting to distinguish between futile exercises and real questions, and I would like to suggest that latent homology be moved into the 'futile exercise' class because, in my opinion, the admission of latent homology into the general category of phylogenetic homology inevitably creates terrible confusion when one tries to use homologies to infer phylogenetic relationships. However, I would like to know other people's opinion on this, as it may depend on whether people have systematic or developmental perspectives on biology.

Meyer: I would like to address this not from a systematic but from a developmental point of view. Latent homology is an interesting concept. It is not futile because all one needs to do is keep separate the homology of the product produced by the developmental process and the homology of the process itself, or the underlying genes that produce molecules which carry out these processes. It is possible to recognize dissociations between homologies at different levels of biological organization, and as long as you're willing to agree that it is possible, it is no longer a futile exercise.

Lacalli: But suppose you have a lineage of organisms where the same gene expression patterns occur in the same part of the body, associated with a particular structure in some instances, but no structure in others. Then, in my view, you have a practical way of mapping the absence of that structure, which is a useful thing to be able to do. In addition to molecular homology and structural homology, you have something in-between, which I like to call regional homology: a tissue domain or region of the body that has certain identifiable characteristics, usually defined in molecular terms, but not always, which cannot otherwise be identified by reliable structural landmarks.

Maynard Smith: What interests me about this is that if the adult structures were present a million years ago, for example, but are now absent, what has been maintaining the machinery for making them?

Meyer: This is also of interest to David Wake. He's interested in homology as a concept that explains biological diversity, and if we use homology as a piece of evidence for evolution we also need to ask what the other pieces of evidence are. Homoplasy is an important piece of evidence that evolution occurs, but the question is, what is more important and more prevalent? Latent homology as a measure of retention of genetic potentiality, i.e. the capability to switch back on X or Y, is an important evolutionary concept. This is a mechanistic process-oriented issue, and in some ways the distinction between homology and homoplasy is futile.

Maynard Smith: Does David Wake discuss examples of homoplasy at the morphological level?

Wilkins: No, he refers to it but doesn't give any examples. One of the classic examples, however, is the development of the carnivorous habit in certain marsupials versus 'proper' carnivores.

Greene: The issue here is not marsupials versus carnivores. This is an ancient split and almost everything involved is homoplasic. What fascinates David Wake, and what fascinates me in behavioural studies, is that when you become familiar with a particular clade—such that you know the morphology, or in my case the defensive behaviour, of every species—and you do a cladistic analysis, you find that there is a lot of homoplasy, but what is bothersome is that homoplasy is hierarchically arranged, so that only above a certain node do you find these recurrent structures or behaviours.

Maynard Smith: What they have in common is the capability of evolving a particular structure.

Greene: In the case of defensive displays in snakes there is even evidence, admittedly slight, that all that is changing is the thresholds for expression.

Wagner: It could be that the problem is not having a proper definition of what the character is, i.e. the homoplasic characters could just be states of a character that were acquired at a particular node. Pre-digits in amphibians (prepollex, prehallex) can be present or absent—e.g. they are prevalent in anurans but do not usually occur in urodeles—and they are always a product of a pre-axial column of developing skeletal elements independent of the digital arch. We have a specimen in *Triturus* which has a pre-digit with three phalanges. It is possible that what is happening is that this is just a character state of this pre-axial series of skeletal elements that can have these terminal elements, but it always has the radiale or tibiale and element Y. Therefore, the actual unit may be the pre-axial series of skeletal elements that can exist with or without a pre-digit. There is a deeper character (the pre-axial series) which may have different character states, i.e. presence or absence of the pre-digit.

Galis: There is also the work of Michael Bell (1988) on sticklebacks. Their predatory armour has been lost in the absence of predators independently many times, but the order in which these elements disappear is usually the same.

Tautz: John Brookfield (1992) and Martin Nowak et al (1997) have calculated that it takes millions of years to lose a gene. Therefore, following speciation events, which occur within a few hundred thousand years, the species will retain the same sets of genes. If the species then hybridize later on the genes will be muddled. Therefore, it is important to know the age of the species you are studying.

Rudolf Raff: We have shown that apparently silenced genes can be maintained over long periods if the gene is functioning in some other process. If not being so used, silenced genes can persist for up to 6×10^6 years, depending on gene size and base distribution and deletion rates within a lineage. Given that the rate of speciation is more rapid than this, characters should be able to switch on and off following speciation events in ways that are alarming for most people doing systematics, i.e. a potent source of homoplasy.

Wray: It is clear that many developmental regulatory genes have multiple functions at different times in the life cycle of the organism. Therefore, the genes are not going to disappear. What changes are the connections they make with other genes and the processes they regulate.

Wagner: It is also possible that our biochemical classifications of gene products and enzymes are not entirely meaningful. For example, the apparent absence of chitin synthetase in vertebrates might be a perceptual problem because the hyaluronic acid synthases are closely related to chitin synthetases. Therefore, we need to be prepared to redefine the boundaries of characters and the definitions of genes once we learn more about their functions.

Roth: We've been talking about the preservation of genes, and their switching on and off, but one could also imagine characters that are apparent at an epigenetic level, requiring tissue interactions or gene product interactions in an entire constellation or network. During evolution such characters could turn on and off with substitution or activation of different sets of genes. It is more difficult to think about the inheritance and dynamics of that kind of evolution, but latent homology at this level also requires some thought.

Hall: It is surprising how few of these regulatory switching genes there are.

Tautz: There may be relatively few regulatory switching genes, but these can act in different combinations. Thus, evolution can probably play around with any number of possible regulatory switches.

Akam: For the purposes of this discussion, the 'genes' we are talking about are really the modules of enhancers. For example, proteins encoded by the *achaete/scute* gene complex are needed to position each bristle on the surface of a fly, but for each bristle or group of bristles these genes are regulated independently by a different regulatory module within the complex, which spans more than 100 kb. The number of such regulatory genes may be small, but each may have many independent regulatory inputs acting through different modules.

Wray: Sir Gavin de Beer (1971) described latent homology as independent appearances in cases where there is an underlying propensity for the appearance of that structure, but the structure itself was not in the ancestor. This relates to Harry Greene's point about structures only appearing above a certain node, because that would be the point at which the propensity appeared.

Striedter: Once you bring up the issue of potentialities, then you have to do experiments to find out what those potentialities are. In the case of behavioural homology, the ability to respond to a certain stimulus can be the character that has been maintained. The organisms may not normally encounter the stimuli necessary to exhibit the behaviour, but if you put them in a situation where they encounter the stimuli that their ancestors encountered, then the behaviour can be manifested. Similarly, the threshold to trigger a behaviour in response to a stimulus might be raised or lowered during evolution, leading to the disappearance or sudden appearance of specific behaviours.

Wilkins: I would like to stress this point with something from George Gaylord Simpson's last book, *Fossils and the history of life* (Simpson 1983). Simpson illustrated the difference between parallel evolution and convergent evolution. He said that we can think of parallel evolution as involving a common pool of genetic potential and then a subsequent divergence of lineages. The evocation of that genetic potential would be similar in those different lineages to give parallel evolution. Convergent evolution, in contrast, involves lineages that did not share any recent common ancestry but in which the genetic changes yield similar phenotypic outcomes under the influence of natural selection. And he gave an interesting example of parallel evolution in the true wolf *Canis lupus*, the Tasmanian wolf *Thylacinus cynocephalus*, which is a marsupial, and the borhyaenid *Prothylacinus patogonicus*, which is also a marsupial. He points out that the jaw and tooth structures of these animals are similar at a certain level, yet the estimate he gives for the divergence between the marsupial and placental lines is 100 million years, whereas that for the divergence between the South American borhyaenid and the Tasmanian wolf is about 60 million years. The interesting point is that these latter two animals evolved from a primitive marsupial that was not carnivorous and did not have jaw structures and teeth like either of these. This is therefore an example of genetic potential that has been evoked in two separate lineages that shared a common ancestry. Perhaps one doesn't want to go quite so far in explaining resemblance of the jaw and tooth structure in the true wolf to the marsupial carnivores, but it is valid for these two marsupial lineages, and it illustrates over what sort of time potential it can be maintained and then evoked in similar ways.

Greene: With respect to marsupials there are cases within a lineage where the same structure is made more than once in the same way because the starting contingencies/constraints were the same. There is a difference between this and something that is literally the same but is re-expressed over and over again. I suspect that these marsupial carnivores are an example of the former, but there are many examples of the latter, and it is the latter that is making us talk about latent homology rather than the former.

Wray: There may be a continuum between these two situations, so rather than trying to worry too much about whether something is convergent or parallel evolution, perhaps we should recognize that in some cases genetic systems will be only a single mutation away from evoking a latent trait, whereas other cases will require quite a lot of reshuffling of machinery.

Greene: It would be exciting to understand how this array of possibilities is produced.

Meyer: The distinction between parallel evolution and convergent evolution is blurred because it involves determining how recent the recent common ancestry is and how similar the similar developmental mechanisms are. One could call it convergent evolution in the case of eutherians/marsupials, but what about different orders of mammals? For example, are the incisors of guinea-pigs and rodents the result of parallel or convergent evolution?

Wray: I don't think you can draw a line between them, it's just a matter of degree.

Hall: One can say the same for homology and homoplasy. If you go far enough back you will find a common ancestor that is homoplasic.

Rudolf Raff: It is possible to have them both together. For example, consider a developmental field of a certain size that produces a certain structure, if the size of that field is increased then duplicated structures are formed as a consequence of the way the field was built. However, if two related organisms independently increase the size of the field by two different mechanisms to produce the 'same' duplication, is it parallelism or convergence?

Roth: It's a matter of character delineation. The *potential* of the field can be homologously the *same* in two organisms, while the size of the field is a distinct characteristic, which in this example changes independently.

Wagner: The danger is that we won't recognize true innovations and that we will lose sight of important biological phenomena.

Hall: Exactly. The problem is where do we set the boundary?

Akam: We recently stumbled on a case that illuminates how characters may appear and disappear repeatedly. The legs of flies sometimes carry a large bristle at the tip of the tibia—the apical bristle. This bristle can appear on any of the legs, and is a useful taxonomic character, both within and between different

Dipteran families. Its formation has clearly been switched on and off repeatedly in the evolution of flies. My colleague Marion Rozowski has been studying the control of this bristle in *Drosophila*, where it appears on the second but not the third leg. She has found that the cell which will make the apical bristle is specified in both the second and the third leg, and begins to differentiate in both. Then the precursor cells in the third leg disappear—we do not know whether they die, or revert to an epidermal fate. Why is this bristle specified and then repressed? Probably because the same genetic routine is being used for the early steps in both legs. Because the bristle is still being specified in both legs, it is easy to see why it might reappear on the third leg in many independent lineages. When it does, is this homoplasy or homology?

Wagner: Gerd Müller pointed out a few years ago that in all cases of experimentally inducible atavisms there are developmental rudiments that are homologous to developmental rudiments in ancestors that had the structure (Müller 1991). The retention of this embryological rudiment is the basis of the continuity between the plesiomorphic character and the atavistic character.

Striedter: I would also like to mention iterative homology. People often argue that this is left out of the phylogenetic concept of homology. The machinery for making a bristle may have evolved at one point in phylogeny and may then be expressed in multiple places on the body of organisms that possess this machinery. Individual bristles in different places would be an instance of iterative homology. However, the developmental machinery for making bristles has a single phylogenetic origin and has been maintained continuously. Therefore, in my opinion, iterative homology is the phylogenetic homology of a developmental machinery that can be expressed in multiple places.

Wilkins: On the other hand, much of the molecular machinery of insect wings and tetrapod limbs is the same, yet we can be quite sure that they evolved quite separately. There was no common ancestral limb structure.

Hall: But are they homologous as appendages?

Wagner: One has to be clear which unit one is studying. Secondary axes of appendages are modules that appear at a certain stage of development and at a certain stage of evolution, so it is a matter of character delineation.

Hall: So the answer of whether they are homologous as appendages might be different to whether they are homologous at the level of similar, homologous genes.

Akam: The only characteristic I'm aware of that specifically unites appendages and does not include any other patterning systems is the expression of *distalless*.

Wray: Like all of these regulatory genes, *distalless* has multiple roles. An additional molecular similarity between insect wings and tetrapod limbs is the expression of *fringe*, which is an edge marker of dorsoventral patterning.

Akam: Are you suggesting that molecular mechanism is seen predominantly in limbs?

Wray: It has been suggested.

Akam: And *wingless* has also been suggested to be a significant shared characteristic of limbs, but many if not all organs are made with *wingless*.

Wagner: One way to look at this is that insect wings and tetrapod limbs not only share genes, but also mechanisms that bind together the capability for proximodistal outgrowth, and anteroposterior and dorsoventral polarization. This machinery characterizes many secondary axes but not all of them.

Müller: I agree, but it's useless to speak of homology at that level because then everything is homologous with everything.

Wagner: No, because there are appendages that don't have this property.

Galis: But what about the primary axis?

Wagner: The patterning there is completely different.

Tautz: At the moment we cannot make any statements because we don't yet have a systematic screen for limb mutants; we are currently relying only on chance observations. In the meantime we should keep an open mind.

Rudolf Raff: I agree. Some of these studies on developmental genetics are still at an early stage, and we don't have the full story. Once someone discovers a new gene they look for it everywhere, so it's not surprising that we find genes such as *wingless* doing jobs that we expect it might be doing in other organisms, based on what it is doing in, say, *Drosophila*, but we don't know all the genes involved in development.

Abouheif: Many members of the *wnt* gene family and *wingless* have different roles in specifying structures in *Drosophila*, mice and chicks.

Akam: If the entire network of anteroposterior and dorsoventral patterning interactions is conserved between vertebrate and insect limbs, I would be happy to say that they are homologous entities, but this is not the case.

Wray: Another important point is that if genes such as *wingless* are expressed at many places and times during development, we have to consider the possibility of chance association. The only way to work this out is not to focus our attention on trying to prove that the genes we know about are expressed everywhere, but rather to find out whether there are so many differences that we have to reject the hypothesis of similarity due to common descent and similarity to convergence.

Meyer: Another problem is that some organisms have several members of the *wingless* gene family, so which one is being compared to the single copy in *Drosophila*?

Hall: And the genes can have different roles in different systems. For example, *dpp* in *Drosophila* is involved in dorsoventral patterning in the nervous system, limb development and mesoderm induction. Clearly, this is a molecule that organisms have found useful, and they have therefore used it over and over again in completely different contexts.

Akam: Eric Davidson might argue that the homology in this case is the invention of mechanisms to pattern fields of cells, as opposed to single-cell interactions. This is a synapomorphy of large metazoa because you can't make a big metazoan unless you can pattern large fields of cells. All of these diffusible signals and gradient morphogens may have been essential at that basal stage.

Carroll: You said that large metazoans would have had these field molecules, and yet before the Cambrian explosion there were presumably a large number of small metazoan lineages that lacked them. They would have had to develop or make use of these molecules, and if they all did this separately what is the point of calling this a homologous situation?

Akam: It's clear that the signalling molecules themselves originated before the diversification of the phyla.

Carroll: The molecules are homologous but what they do is not, and the lineages they do it in are different.

Wray: The signalling molecules are a good example because there are many inductive interactions during development. In fact, there are many more inductive interactions during development than there are signalling molecules, although we can't yet put any numbers on these. This means that these signals have to be used over and over again.

Meyer: Evolution is going to work that way; it's going to be lazy, and it's going reuse the same genes, the same networks and the same genetic potentialities over and over again in different contexts.

Akam: I would like to challenge that. Clearly, the same cellular machinery is being used, i.e. the Wnt signal transduction system, but it is still an open question as to how often an entire genetic network is recruited. The genetic network for the immune response has been recruited for dorsoventral patterning in the *Drosophila* embryo, but that is probably a rare case. It is more likely that when a new structure is being built, cell identity molecules are recruited piecemeal. The link between transcription factors and target genes may be quite labile in evolution.

Lacalli: You refer to Eric Davidson's idea about the problem of controlling development in large tissue domains, but in the pre-Cambrian there were animals as big as doormats. I find it difficult to see why we should restrict our thinking about ancestral metazoans to tiny animals, when some of our earliest fossils are enormous.

Wray: We're not sure that Ediacaran organisms like *Dickinsonia* and *Pteridinium* were animals.

Lacalli: I agree there is a problem interpreting these fossils; what is the current consensus in your view?

Rudolf Raff: They are peculiar creatures. They have fractal structures, so it is not clear that they are metazoans at all.

Lacalli: But it is still possible that *Dickinsonia*, for example, is some sort of giant polychaete, though it looks like it has been flattened by a steamroller. You also have organisms like *Spriggina*, which look more like a modern metazoan, and yet are still several centimetres in length. This is considerably larger than the tiniest of modern flatworms, which are the usual models for ancestral bilateral metazoans.

Rudolf Raff: Animals that are only a few millimetres in length are still large in terms of setting up mechanisms for signalling between cells and tissues.

Lacalli: I agree that there could be a big difference between a fauna of tiny animals, no more than 200 μm, or of small animals 2 mm or even 2 cm long. Achieving the great size of 2 cm could have been major event in metazoan evolution.

Akam: Put in a slightly different way, there are mechanisms that allow cells to be specified as different from one another based purely on the segregation of determinants within cells and cells physically touching one another. Eric Davidson thinks these are the key mechanisms operating in what he thinks of as ancestral development. Signals operating at a distance to generate pattern across many cell diameters would be bolted onto to the ancestral mechanism.

Lacalli: I question whether this is a convincing argument.

Hall: Organisms do compartmentalize so that signals are passed within a compartmentalized population of cells and to other populations of cells. Those are the two ways to generate diversity.

Lacalli: We haven't yet discussed what some of these unknown common ancestors may have been like. In some evolutionary circles this is a bad thing to do, but in terms of looking at developmental phenomena and trying to rationalize how they may have evolved I would argue that it is necessary to use one's imagination to think about what sort of organisms they might have been.

Wagner: This is the step from phylogeny construction to character evolution. At the molecular level it is possible to reconstruct the ancestral molecules, synthesize them and test them, as for instance in the case of lysozymes, so with better technology we may be able to do this in the future at the developmental level.

Hall: The embryos of many vertebrates that have lost their limbs, e.g. whales which have lost their hindlimbs, and some snakes and legless lizards, have limb buds. They all carry these limb buds through to more or less the same stage of development, and then they regress. Therefore, there has clearly been a conservation of some basic patterning information that is used in other parts of the organism as well.

Akam: Is there an explanation for why they have kept the limb buds?

Hall: The traditional explanation is that the loss of the entire limb bud would entail the concomitant loss of much of the information specifying the basic patterning of the body axis, so that large areas of the embryo would be disrupted. But I'm not sure if this is a satisfactory explanation.

Müller: It cannot be fully satisfactory because there are also instances where the limb buds have been lost completely, such as in caecilians. And the axial patterning of these organisms seems quite ok.

Galis: The gene activity involved in specifying the limb is not necessarily involved in the patterning of the vertebrate body axis. The regulation of the genes during the development of these different structures appears to be independent of each other (Cohn et al 1997, Zakany et al 1997)

Peter Holland: When a structure is lost in evolution why does the defect in a pathway often occur late in development?

Akam: The changes do not have to be late. There are other examples where the defect occurs early. In the case of one bristle that Marion Rozowski looked at, she found that the bristle primordium is missing, suggesting that it is repressed at an early developmental stage. Evolution seems to have taken different routes in different cases; some routes may be amenable to reversion and some may not. I'm sure that the patterns of character change are consistent with this.

Tautz: In *Drosophila* we know that the position of the limbs is determined early in development by the crossing of *wingless* and *dpp* signals, and this is before the *Hox* genes become active. It seems almost impossible that a defect at this stage would result in a limb not being formed.

Müller: This discussion seems to have reached a point that reflects the general confusion in the literature on homology. And the confusion arises because we are constantly mixing up levels, such as defining homologies of the anatomical level by shared developmental or genetic programmes. We also cannot say something is homologous at the phenotypic level only because it is used for the same behaviour. We can homologize anatomical characters at the level of anatomical construction. We can also homologize behaviour, but only at the level of behaviour. We can homologize genes, and maybe even gene regulatory networks, but only at the level of the genome. And we can homologize developmental pathways, but only at the developmental level. However, we must not define homology at one level via mechanisms that belong to a different level of analysis.

Peter Holland: I would like to defend the opposing point of view. It is not really that we 'mix up' levels of homology, and we are not jumping out of control between them. These various levels of homology are not independently shifting planes. There may be a dependence between levels. That dependence may not always go in both directions, but clearly there is communication between the levels.

Response by David Wake

My omission of a precise definition of partial homology led to discussion that suggests that I should make my perspective clear. Because evolution is a continuous process, I believe that homology can only ever be partial, in any real sense. A particular *Hox* gene is identified as such by some general criterion, but when the gene is examined in detail in two taxa, even closely related taxa, there will doubtless be found to be some small differences in base pair composition, for example. By the general criterion, the two genes are exact homologues, but at the next hierarchical level they are only partial homologues. Vertebrae are homologous as vertebrae, defined again by some general criterion (e.g. those bones forming the axial skeletons of osteichthyans), but they are not identical in structure even when comparing rather close relatives. Then there is the issue of strict partial homology, as occurs in cases of exon shuffling, as discussed by Hillis (1994).

There is also some question as to what elements of my critique were the most important. I will clarify my view. I certainly do not want to abandon the concept of homology; rather, I want to stop the fruitless discussion of what it is! Modern developmental genetics has given us the tools to create a new kind of evolutionary biology, one that is focused on the production of phenotypes and on the factors and processes that enable biological diversification to proceed. I want to get on with it and to leave behind debates that started when biologists really did not have sufficient biological knowledge to appreciate the causes of biological similarity and when they did not yet understand that Darwin was right in his view that there is one genealogy for all of life. Common ancestry is all there is to homology. Accordingly, a central theme of my chapter is that homology has necessarily become an abstract concept. When I say that the concept of homology has become reified I mean that the abstraction is being converted into a concrete material object by those who continue to debate what the essence of homology might be.

Axel Meyer has correctly expressed my view that homology is fundamental to questions of biological diversity. Homoplasy is an alternative perspective on homology, and when we can identify a phenomenon as latent homology we begin to approach an understanding of how homoplasy relates to homology on the one hand and to the production of diversity on the other. As Jacob (1977) so clearly enunciated, evolution works with materials on hand. The fact that genes, gene products, cellular and developmental processes, and even parts of organisms that we call homologues hang around to be used again and again bears testimony to homology, but especially to the endless possibilities that appear to us as homoplasies. It is understanding these processes that will lead to full appreciation of what I called the real problems of modern biology (for a useful perspective on these issues see Kirschner & Gerhart 1998).

References

Bell MA 1988 Stickleback fishes: bridging the gap between population biology and paleobiology. Trends Ecol Evol 3:320–325

Brookfield J 1992 Evolutionary genetics: can genes be truly redundant? Curr Biol 2:553–554
Cohn MJ, Patel K, Krumlauf R, Wilkinson DG, Clarke JDW, Tickle C 1997 *Hox9* genes and vertebrate limb specification. Nature 387:97–101
de Beer GR 1971 Homology: an unsolved problem. Oxford University Press, Oxford (Oxford Biol Readers 11)
Hillis DM 1994 Homology in molecular biology. In: Hall BK (ed) Homology: the hierarchical basis of comparative biology. Academic Press, San Diego, CA, p 339–368
Jacob F 1977 Evolution and tinkering. Science 196:1161–1166
Kirschner M, Gerhart J 1998 Evolvability. Proc Natl Acad Sci USA 95:8420–8427
Müller GB 1991 Experimental strategies in evolutionary embryology. Am Zool 31:605–615
Nowak MA, Boerlijst MC, Cooke J, Maynard Smith J 1997 Evolution of genetic redundancy. Nature 388:167–171
Sattler R 1994 Homology, homeosis, and process morphology in plants. In: Hall BK (ed). Homology: the hierarchical basis of comparative biology. Academic Press, San Diego, CA, p 424–475
Simpson GG 1983 Fossils and the history of life. WH Freeman, San Francisco
Wagner GP 1996 Homologs, natural kinds and the evolution of modularity. Am Zool 36:36–43
Wagner GP 1999 A research programme for testing the biological homology concept. In: Homology. Wiley, Chichester (Novartis Found Symp 222) p 125–140
Zakany J, Gérard M, Favier B, Duboule D 1997 Deletion of a *HoxD* enhancer induces transcriptional heterochrony leading to transposition of the sacrum. EMBO J 16:4393–4402

Homology among divergent Paleozoic tetrapod clades

Robert L. Carroll

Redpath Museum, McGill University, 859 Sherbrooke St. West, Montreal, Canada H3A 2K6

Abstract. A stringent definition of homology is necessary to establish phylogenetic relationships among Paleozoic amphibians. Many derived characters exhibited by divergent clades of Carboniferous lepospondyls resemble those achieved convergently among Cenozoic squamates that have elongate bodies and reduced limbs, and by lineages of modern amphibians that have undergone miniaturization. Incongruent character distribution, poorly resolved cladograms and functionally improbable character transformations determined by phylogenetic analysis suggest that convergence was also common among Paleozoic amphibians with a skull length under 3 cm, including lepospondyls, early amniotes and the putative ancestors of modern amphibians. For this reason, it is injudicious to equate apparent synapomorphy (perceived common presence of a particular derived character in two putative sister-taxa) with strict homology of phylogenetic origin. Identification of homology by the similarity of structure, anatomical position and pattern of development is insufficient to establish the synapomorphy of bone and limb loss or precocial ossification of vertebral centra, which are common among small Paleozoic amphibians. The only way in which synapomorphies can be established definitively is through the discovery and recognition of the trait in question in basal members of each of the clades under study, and in their immediate common ancestors.

1999 Homology. Wiley, Chichester (Novartis Foundation Symposium 222) p 47–64

The fish–amphibian transition

One of the most significant events in the history of life was the emergence of terrestrial vertebrates in the Late Devonian, approximately 365 million years ago. This transition has been extensively studied by Clack (1999), Coates (1996) and Jarvik (1996). Their discoveries show a series of significant changes in the dermal skull, braincase and appendicular skeleton occurring in a mosaic pattern over a period of 10–15 million years, leading from obligatorily aquatic fish to animals with limbs that indicate at least facultative terrestrial locomotion.

While there were major changes in structures associated with support and locomotion on land, other aspects of the earliest terrestrial vertebrates retained the pattern of their fish ancestors. Reproduction remained tied to the water, as is documented by the many gilled larvae known from the Carboniferous and Early Permian. The basic pattern of the skull bones closely resembles that of their fish ancestors, as do their vertebrae, which were composed of several separate elements.

Of particular importance regarding their way of life was the retention of body dimensions and proportions comparable to those of their fish ancestors, with skulls 12–20 cm in length and skull–trunk lengths of 40–60 cm. The earliest known amphibians had long tails, with substantial caudal fins. The retained capacity for effective aquatic locomotion as well as their cranial and dental similarities suggest that the early amphibians may have fed on prey similar to that of their osteolepiform ancestors, which probably consisted primarily of relatively large fish and perhaps aquatic arthropods.

Animals of comparable body size, proportions and vertebral structure (collectively called labyrinthodonts because of the infolding of the dentine of their teeth) dominated the land throughout the Carboniferous, and remained common and diverse into the Early Mesozoic. The labyrinthodonts were not, however, the only group of Paleozoic amphibians. While early labyrinthodonts represent the first stage in adapting to life on land, a second, major shift in anatomy and way of life occurred during the Carboniferous that eventually led to both the modern amphibian orders and the amniotes, including the ancestors of modern reptiles, birds and mammals.

Miniaturization in early tetrapods

In addition to the large, basically conservative labyrinthodonts, there appeared, beginning in the Early Carboniferous, several lineages of much smaller but highly ossified animals: some presaging the origin of amniotes; and others belonging to a diverse but little appreciated assemblage termed the lepospondyls (Fig. 1 and Table 1). It is these animals that I will be discussing, for they raise a number of questions regarding the recognition of homology and its use in establishing phylogenetic relationships.

The first question is why small size would, fairly suddenly, take on a strong selective advantage among early tetrapods. One possible explanation is the appearance of a new food source that was both small and common. This was presumably terrestrial arthropods, especially insects, which had been present since the Lower Devonian but underwent an explosive radiation during the Carboniferous, leading to nearly 90 families by the Lower Pennsylvanian. Hanken (1993) attributed miniaturization among modern salamanders to precocial ossification that terminated growth at small size. This also explains the high degree of ossification of even the smallest lepospondyls and early amniotes.

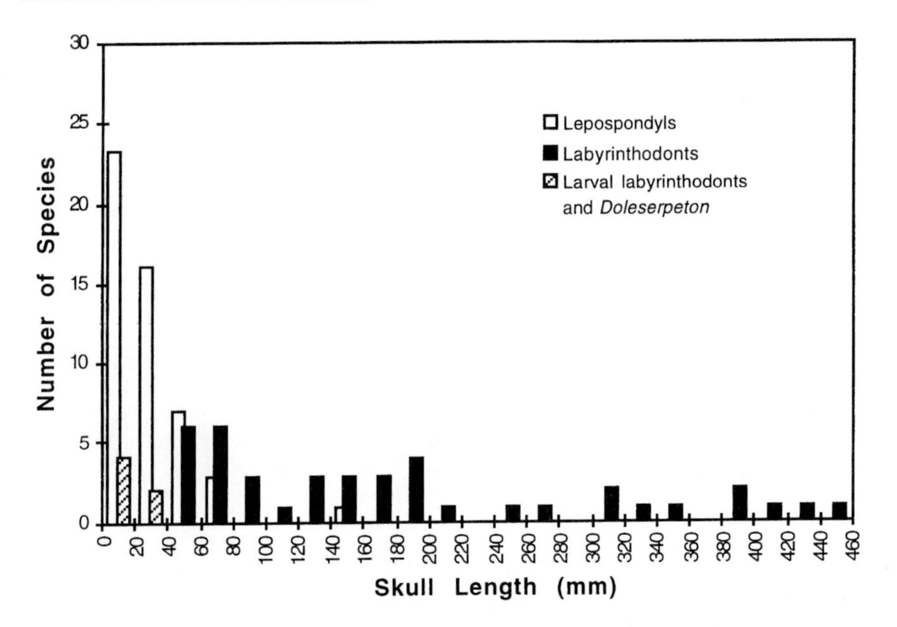

FIG. 1. Size distribution of cranial length in Paleozoic amphibians. Open bars, lepospondyls; solid bars, adult labyrinthodonts; hatched bars, larval, possibly neotenic labyrinthodonts and *Doleserpeton*, which are strongly supported candidates for the sister-taxa of the living amphibian orders, particularly the frogs. Data on which this graph was based appear in Table 1. In contrast with lepospondyls, neotenic labyrinthodonts retained the slow developmental pattern of the vertebrae common to larger labyrinthodonts. On the other hand, *Doleserpeton* (Bolt 1991) co-ossified the centra and neural arches at small body size, and evolved a single unit atlas vertebra resembling that of most lepospondyls and the modern amphibian orders.

Problems of homology among lepospondyls

Small size brought with it not simply a capacity to feed more effectively on small prey, but also a host of other skeletal changes that resulted in much more anatomical diversity than that seen among the labyrinthodonts and numerous problems involving homology. These become of utmost importance in regard to establishing the relationships of these small animals, which may include the ancestors of all living terrestrial vertebrates.

Laurin & Reisz (1997) published the first cladogram (Fig. 2) based on extensive phylogenetic analysis of all major groups of Paleozoic amphibians, early amniotes and representatives of the modern amphibian orders. Their union of amniotes and modern amphibians in a monophyletic assemblage above the level of the labyrinthodonts is different from the pattern of evolution previously accepted by most paleontologists, who had argued that amniotes and modern amphibians had

TABLE 1 Skull lengths of Paleozoic tetrapods

Labyrinthodonts	*Skull length (mm)*	*Lepospondyls*	*Skull length (mm)*
Primitive genera		Microsauria	
Acanthostega	127	*Tuditanus*	23
Crassigyrinus	316	*Asaphestera*	42
Ichthyostega	250	*Crinodon*	37
Anthracosauria		*Boii*	23
Eoherpeton	156	*Hapsidopareion*	12
Proterogyrinus	219	*Goreville* sp.	17
Palaeoherpeton	170	*Llistrofus*	23
Eogyrinus	426	*Saxonerpeton*	11
Archeria	189	*Pantylus*	67
Anthracosaurus	384	*Pariotichus*	27
Neopteroplax	400	*Leiocephalikon*	22
Leptophractus	118	*Cardiocephalus*	20
Temnospondyli		*Euryodus*	38
Dendrerpeton	92	*Pelodosotis*	58
Greererpeton	156	*Micraroter*	62
Neldasaurus	173	*Goniorhynchus*	18
Tersomius	60	*Microbrachis*	23
Phonerpeton	91	*Hyloplesion*	17
Broiliellus	67	*Brachystelechus*	9
Brevidorsum	74	*Odonterpeton*	7
Conjunctio	84	*Tribecaton*	26
Amphibamus	68	*Rhynchonkos*	18
Eryops	443	*Batropetes*	10
Caerorhachis	53	*Carrolla*	18
Edops	520	*Quasicaecilia*	15
Cochleosaurus	156	*Utaherpeton*	10
Chenoprosopus	309	Nectridea	
Balanerpeton	44	*Batrachiderpeton*	54
Capetus	269	*Diceratosaurus*	20
Acanthostomatops	79	*Diploceraspis*	30
Zatrachys	123	*Diplocaulus*	147
Cacops	132	*Scincosaurus*	13
Kamacops	332	*Sauropleura*	25
Micropholis	40	*Ptyonius*	70

TABLE 1 Skull lengths of Paleozoic tetrapods *(Contd.)*

Labyrinthodonts	*Skull length (mm)*	*Lepospondyls*	*Skull length (mm)*
Parioxys	182	*Urocordylus*	40
Trimerorhachis	182	*Keraterpeton*	36
Isodectes	44	*Lepterpeton*	28
Acroplous	66	Lysorophia	
Tupilakosaurus	63	*Brachydectes*	18
Sclerocephalus	194	*Pleuroptyx*	25
Cheliderpeton	161	Adelospondyli	
Platyoposaurus	417	*Adelospondylus*	53
Kunzhukoria	346	*Palaeomolgophis*	28
Doleserpeton	14	*Adelogyrinus*	51
Branchiosauria		*Dolichopareias*	51
Branchiosaurus	16	Aïstopoda	
Micromelerpeton	14	*Lethiscus*	16
Apateon	15	*Phlegethontia*	18
Melanerpeton	32	*Oestocephalus*	50
Schoenfelderpeton	22	Undesignated	
		Acherontiscus	15

Data from Heatwole & Carroll (1999), Carroll et al (1998) and Werneburg (1989).

diverged ultimately from separate groups of labyrinthodonts near the base of the Carboniferous, or even by the Late Devonian (Panchen & Smithson 1988, Lebedev & Coates 1995). The key clearly lies with the phylogenetic position of the lepospondyl amphibians.

The lepospondyls were a diverse assemblage of Carboniferous and Permian amphibians that paralleled the habitus of several groups of small modern tetrapods. Some were elongate and limbless, superficially resembling modern snakes and caecilians, some resembled modern lizards, whereas others covered much of the range of salamander adaptive modes.

Figure 3 shows the skeletons of early representatives of each of the five lepospondyl orders. All are known from nearly complete skeletons, and show significantly divergent structures. The aïstopods were the most highly derived, but the first of these groups to appear in the fossil record, within about 15 million years of the Devonian tetrapods. The earliest known genus, *Lethiscus*, shows 78 vertebrae in articulation, but certainly not including the end of the body. There is no trace of pelvic girdle or rear limb; some elements may be

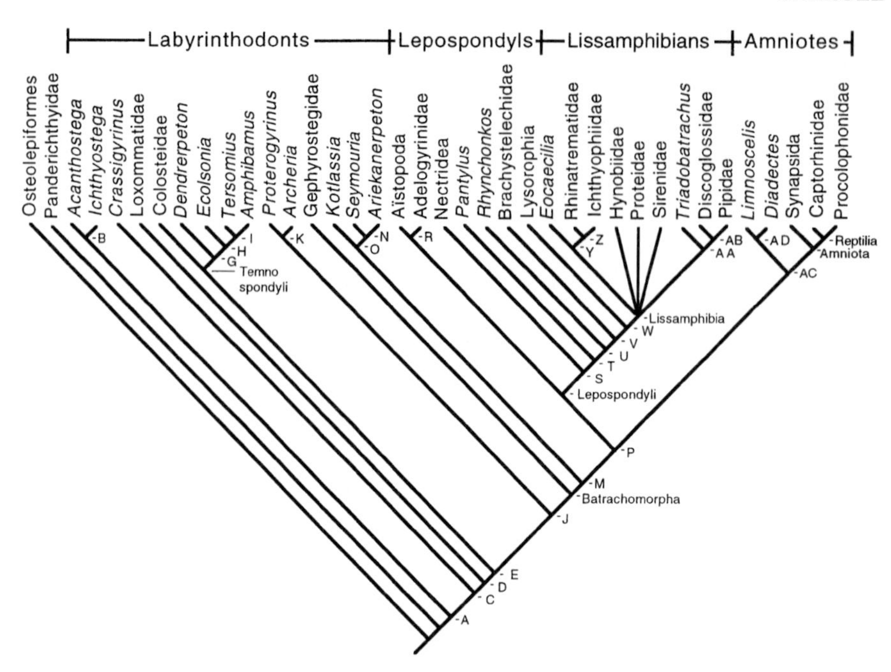

FIG. 2. Cladogram of Paleozoic and modern amphibian groups and early amniotes (from Laurin & Reisz 1997). Amniotes appear as a sister-taxon of lepospondyls, within which are nested representatives of the frogs, salamanders and caecilians. This contrasts strongly with the generally accepted hypothesis that the ultimate ancestors of the amniotes and modern amphibians had diverged from different clades of labyrinthodonts in the Early Carboniferous or even the Late Devonian. Lepospondyls (specifically microsaurs) are otherwise postulated only as a plausible sister-taxon to the caecilians (Milner 1993, Heatwole & Carroll 1999).

remnants of the pectoral girdle, but no forelimb is present. More derived aïstopods have up to 240 vertebrae, but none show a trace of limbs.

The adelospondylids also lack limbs and the pelvic girdle, but have a massive dermal shoulder girdle and a huge hyoid apparatus, a structure that shows no trace of ossification in aïstopods. The lysorophids also have a long vertebral column, with up to 97 presacrals, but retain reduced limbs and girdles over the 40 million years of their evolution. The nectrideans, in contrast, never have more than 26 presacral vertebrae, but possess a long tail with haemal arches fused to the centra and closely resembling the neural arches. Their limbs may be reduced in size and degree of ossification, but are always retained. Microsaurs lack the conspicuous specializations of the previous groups, although one genus stretches the trunk to 44 vertebrae. Some microsaurs are primitive in the retention of intercentra in the trunk vertebrae, whereas all the other groups have lost them.

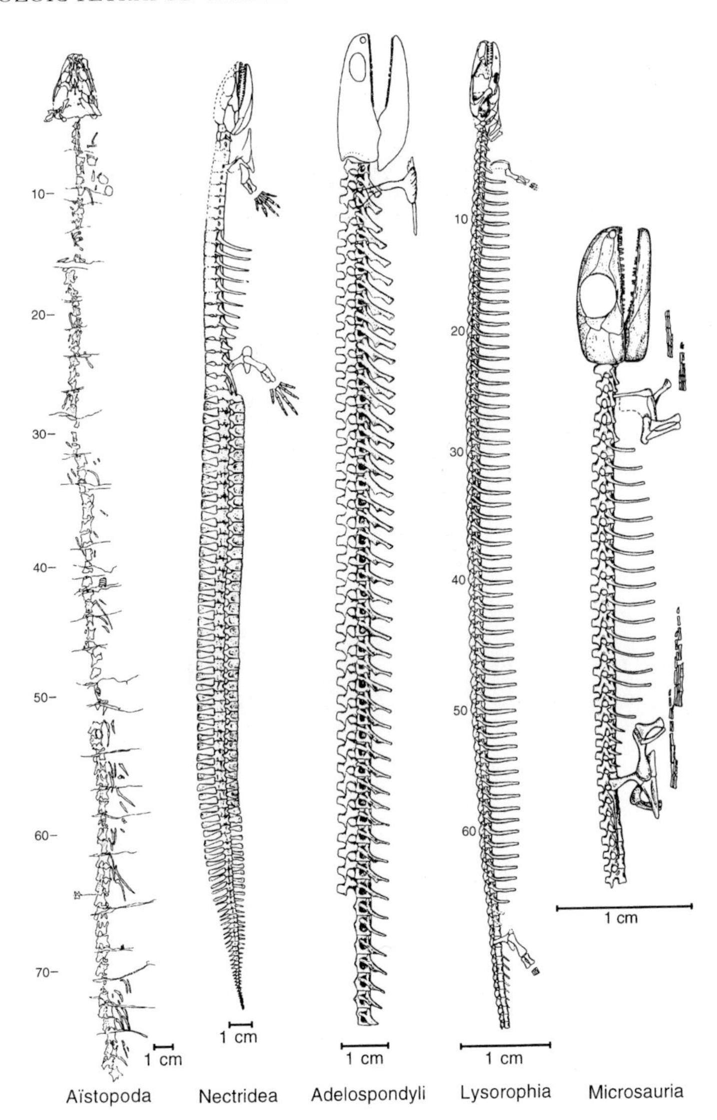

FIG. 3. Reconstructions of the skeletons of the earliest and most primitive adequately known species of the five major clades of lepospondyls. The Aïstopoda is represented by *Lethiscus stocki* from the Mid-Viséan, Lower Carboniferous, of Scotland. The Nectridea is represented by *Urocordylus wandesfordii*, Westphalian A, Upper Carboniferous, Ireland. The Adelospondyli by *Palaeomolgophis scoticus*, Viséan of Scotland, Lower Carboniferous. The lysorophid is *Brachydectes newberryi*, Westphalian D, Upper Carboniferous, Ohio. The microsaur is *Utaherpeton franklini*, lower portion of Upper Carboniferous, Utah. An earlier microsaur has been reported but not fully described from the Chokurian stage of Goreville, Illinois (Schultze & Bolt 1996). Illustrations modified from Carroll et al (1998).

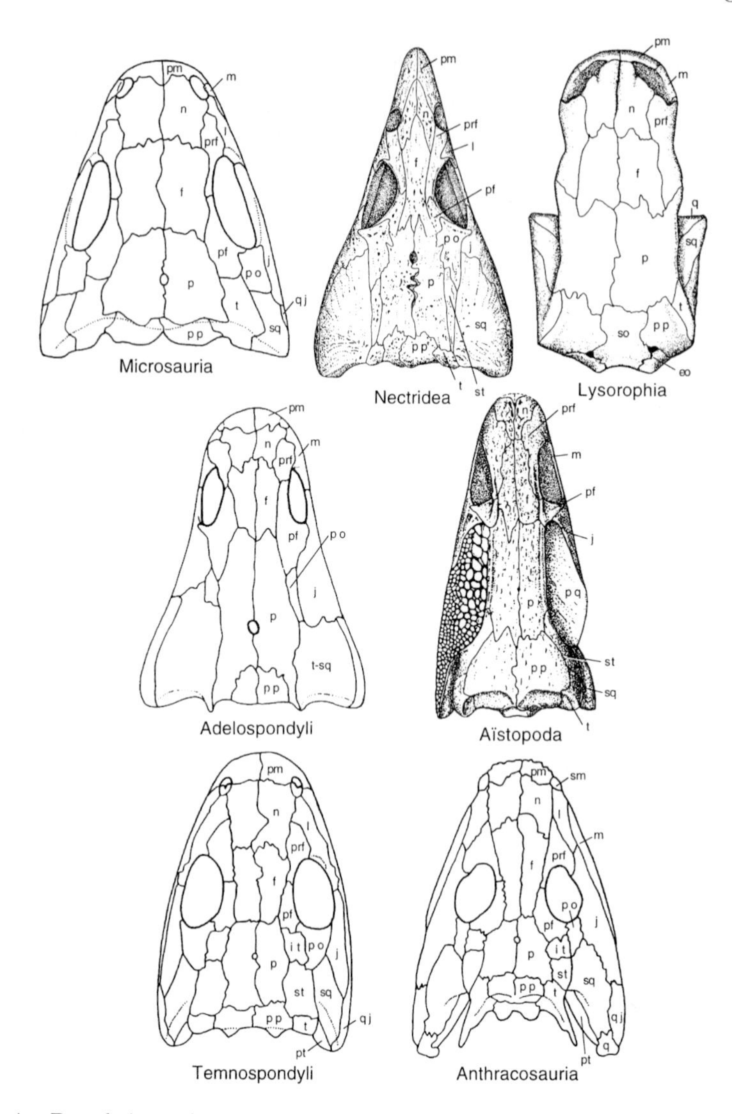

FIG. 4. Dorsal views of skulls of the five major clades of lepospondyls and representatives of the major Permo-Carboniferous labyrinthodont groups. The Microsauria is represented by *Asaphestera intermedia*, Westphalian A, Upper Carboniferous, Nova Scotia. The nectridean is *Sauropleura scalaris*, Westphalian D, Linton, Ohio. The lysorophid is *Brachydectes elongatus*, from the Lower Permian of Texas. The skull differs only in proportions from the Upper Carboniferous *Brachydectes newberryi*. The extensive fenestration of the cheek can be seen in Fig. 3. The adelospondyl is *Adelogyrinus simorhynchus*, Lower Carboniferous of Scotland. Temnospondyls are represented by *Dendrerpeton acadianum* from the Westphalian A of Joggins, Nova Scotia, and the anthracosaur is *Palaeoherpeton decorum*, from the Upper Carboniferous of Great Britain. Both labyrinthodonts retain the intertemporal bone that is missing in all

The skulls of the various lepospondyl orders are also highly distinctive (Fig. 4). The aïstopods have a large opening in the cheek, and a connection between the occiput and first cervical that is unique among tetrapods in having a narrow circular band of bony tissue around a deep pit in both elements. Nectrideans are the most primitive in the pattern of their skull roof, losing only a single bone (the intertemporal) from the pattern in primitive labyrinthodonts. Lysorophids have a large opening extending from the eye to the jaw suspension, which angles sharply forward. Adelospondyls have a solid covering of the back of the skull, but have lost or fused several of the bones present in more primitive tetrapods. Microsaurs typically have a single large bone in the temporal region, called the tabular, incorporating the areas of the supratemporal, intertemporal and tabular of more primitive tetrapods, and reaching both the postorbital and the postfrontal.

The genera shown in Figs 3 & 4 are among the earliest known members of each of these groups, and all of them show nearly every bone in the body. Yet, no member of any of these groups provides obvious evidence of relationships with other Paleozoic tetrapods, or with living amphibians or amniotes. However, all of them do exhibit the following unique derived characters, relative to both labyrinthodonts and their osteolepiform ancestors, which suggest that they shared a common ancestry:

(1) absence of palatal fang and pit pairs;
(2) absence of labyrinthine infolding of the dentine;
(3) jaw articulation at the level of, or anterior to, the occiput;
(4) loss of one or more skull bones present in more primitive tetrapods;
(5) absence of squamosal embayment;
(6) absence of atlas intercentrum;
(7) atlas pleurocentrum, a median, cylindrical element; and
(8) cylindrical trunk centra, occupying the entire segment.

These features, especially the cylindrical nature of the vertebral centra, have suggested the monophyletic origin of the lepospondyls since their recognition in the late 19th century. However, establishing interrelationships among these orders is much more difficult. I have published two phylogenetic analyses of this assemblage, one (Carroll 1995) as part of a global study of all Paleozoic tetrapods, and another (Carroll & Chorn 1995) dealing only with the vertebral characters of

lepospondyls (modified from Carroll et al 1998). eo, exoccipital; f, frontal; it, intertemporal; j, jugal; l, lacrimal; m, maxilla; n, nasal; p, parietal; pf, postfrontal; pm, premaxilla; po, postorbital; pp, postparietal; pq, palatoquadrate; prf, prefrontal; pt, pterygoid; q, quadrate; qj, quadratojugal; sm, septomaxilla; so, supraoccipital; sq, squamosal; st, supratemporal; t, tabular; t-sq, bone occupying position of tabular and squamosal.

the lepospondyls. These yielded different results, but both showed extremely incongruent character distribution, functionally improbable character transformations, and did not retain their resolution more than a few steps beyond the most parsimonious trees. These results could be interpreted as reflecting a pattern of evolution in which the derived characters common to all groups accumulated over a considerable period of time, followed by extremely rapid divergence with few synapomorphies accumulating during the short periods between successive bifurcation of the individual clades. On the other hand, many characters that appear to unite particular clades, and also those that are common to all lepospondyls, may be highly correlated with one another, either as integrated elements of functional complexes, or because they reflect similar responses to changes in size and/or patterns of development. Close functional or developmental correlation of many characters might result in misleading patterns of apparent relationship based on searches for the most parsimonious distribution of all character changes.

Structural and positional homology of trunk elongation and limb loss

The cladograms of Carroll (1995) strongly supported union of the aïstopods and adelospondylids, with the lysorophids as the sister-group of these two clades. The reason is obvious: all are greatly elongate forms, with the girdles and limbs reduced or absent. In an effort to classify all Paleozoic tetrapods, all differences in limb elements were coded as separate traits. Reduction in phalanges, digits and entire limbs was common throughout this assemblage. In groups showing extensive limb reduction or complete loss, these characters would swamp all others. In these three orders, this is also correlated with greatly increased number of vertebrae. Although the high degree of reduction of the appendicular skeleton and elongation of the trunk may be synapomorphies of these three groups, this seems to be contradicted by the very different patterns of the skull, and also by details of the anatomy of the individual vertebrae. In all aïstopods the arch and centra are fused without a trace of suture, in adelospondylids the arch is loosely attached to the centrum and bears a long transverse process. In lysorophids the arches are paired, rather than fused at the midline, and are loosely attached to the centrum.

Furthermore, knowledge of modern lizards and snakes makes one hesitate to accept trunk elongation and limb reduction as indicative of close relationships. Greer (1991) tabulated convergent evolution of trunk elongation and limb reduction a minimum of 62 times in 53 lineages of modern lizards. Gans (1975) discussed the close association of these attributes, which commonly occur in small animals living close to the substrate as in the case of the lepospondyls. Lee (1999) is currently studying the relationships of snakes, amphisbaenids and dibamids with other squamate taxa. Trunk elongation, limb reduction and cranial specializations associated with burrowing result in a host of similarly derived characters that lead to these lineages being allied in a single, monophyletic group if only living species are considered. In contrast, inclusion of data from more primitive fossil taxa shows that these groups evolved separately from quadrupedal ancestors.

The difficulty of judging the patterns of relationship and evolution among lepospondyls alerts us to the problems of recognizing and categorizing homology. Aïstopods, adelospondylids and lysorophids share similarly derived characters that should enable us to establish their relationships, but in what ways are these characters homologous? Based on the recognition of homology through similarity of the structure and relationship of the units, all the bones that make up the limbs and vertebrae in each taxon are homologous with those in other taxa. Similar changes in these bones should also be considered as homologous. Reduction and loss of limb bones, in both living and fossil tetrapods, show similar patterns that imply similar developmental constraints, which accords with the concept of developmental homology (termed biological homology by Wagner 1994). On the other hand, neither the structural nor developmental homology of the derived characters exhibited by these three taxa necessarily demonstrates that they evolved from character states already apparent in their immediate common ancestor, as is required by the phylogenetic concept of homology (Shubin 1994).

The only way in which the phylogenetic homology of limb loss and trunk elongation in these groups can be established is through knowledge of more primitive fossil representatives of each of these groups. There is currently a gap in the fossil record of 15 to 55 million years between the earliest known members of these derived groups and the appearance of early labyrinthodonts in the Upper Devonian that should eventually yield appropriate fossils to solve this problem.

Miniaturization and developmental homology

Even more serious problems are encountered in establishing the unity of the lepospondyls as a group, and their putative close relationships with the modern amphibian orders and the amniotes, as proposed by Laurin & Reisz (1997).

Returning to the list of apparent synapomorphies of the lepospondyls, this looks like a substantial number. Moreover, many of these also apply to both living amphibians and early amniotes. All but number 5 apply to all modern amphibians, and all but 6 and 7 apply to early amniotes. On the other hand, characters 1 through 3 also occur in larval, but not adult labyrinthodonts. The typical, adult labyrinthodont condition is not achieved until the young have reached a skull size in excess of that in most early lepospondyls, early amniotes and putative ancestors of the modern amphibian orders. That is, many of these features could have evolved in the ancestors of all living tetrapod groups simply by truncation of growth at small size. This hypothesis is supported by the presence of at least some of these features in the labyrinthodont character state in especially large lepospondyls and amniotes. Labyrinthine infolding appears in two, very large sister-taxa of early amniotes, in one very large specimen of a nectridean species and in one very large microsaur. Some nectrideans and microsaurs also exhibit posterior extension of the jaw articulation in large individuals, and loss of additional skull bones occurs in especially small microsaurs and nectrideans. The traits that distinguish lepospondyls can be considered structurally and developmentally homologous, but were they phylogenetically homologous? The key to determining phylogenetic relationships is establishing whether or not all these groups shared an immediate common ancestor that was small enough as an adult to retain the juvenile characters into the mature individual.

As in the case of trunk elongation and limb reduction, the number of cranial characters associated with small size is enough to swamp other characters that might suggest separate ancestry, such as the specific patterns of the cranial bones and configuration of the palate.

Although several derived cranial features of lepospondyls and early amniotes can be accounted for simply by truncation of growth, this is not the case for the vertebrae, which show an entirely different pattern and rate of development. The large, primitive labyrinthodont amphibians had markedly prolonged larval stages, during which the vertebrae were slow to ossify (Fig. 5). First the skull, then the limbs, but only in individuals exceeding 10 cm in skull–trunk length are the vertebrae completely ossified. The neural arches ossify first, and then the centra. In contrast, both lepospondyls and early amniotes ossify their vertebrae at much smaller body size, down to about 2 cm in skull–trunk length.

The precocial ossification of the vertebrae presumably resulted from the same factors that led to truncation of growth of the entire skeleton. Ossification at extremely small size is also a probable cause for the simplification of the vertebrae in all the smaller groups. Hall & Miyake (1995) pointed out that a minimum size and/or number of cells are necessary for the establishment of separate centres of ossification. With fewer cells, small centres of mesenchymal condensation will coalesce to form only a single centre of ossification. Hence, most lepospondyls, amniotes and modern amphibians have a single cylindrical centra per segment. Thus, the cylindrical configuration of the centra in all groups, like changes in the skull, may be attributed to small size and not necessarily to common ancestry.

At least one vertebral feature that does support specific relationship, rather than just small size, is the configuration of the atlas–axis complex in early amniotes. This is a multipartite structure comparable with that in early

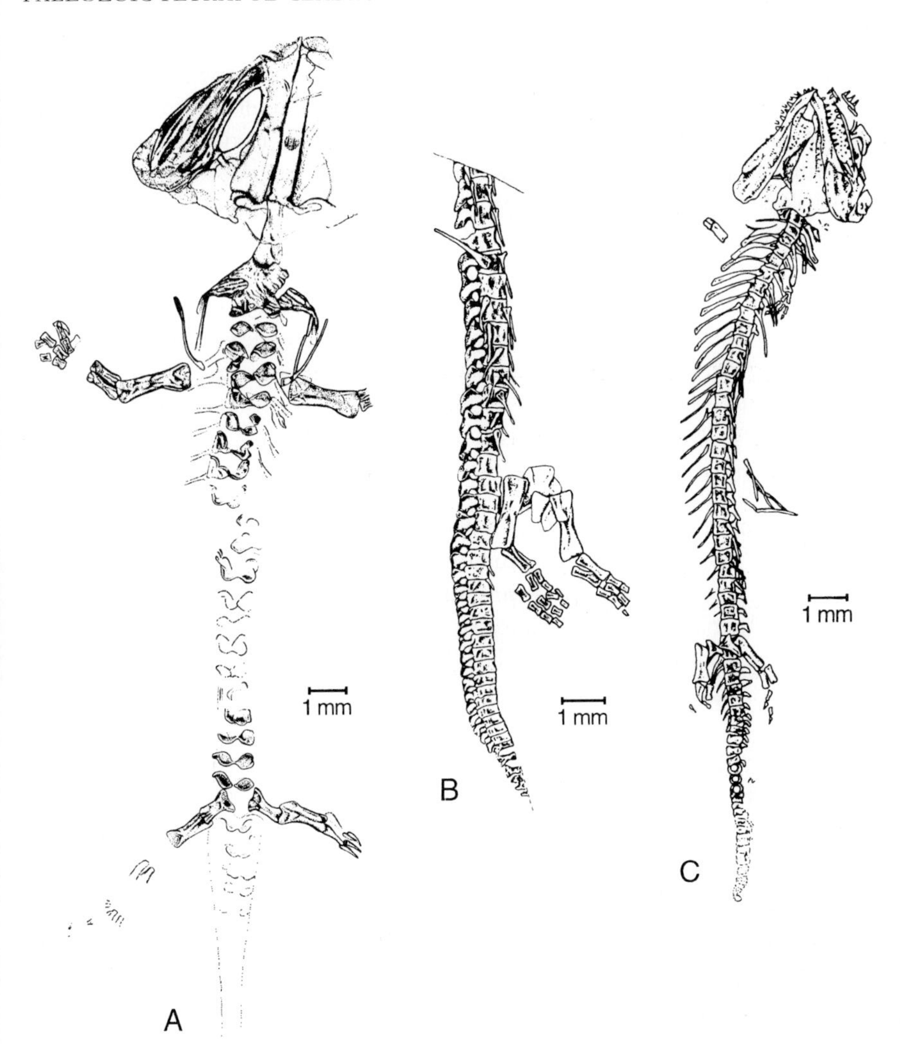

FIG. 5. Paleozoic amphibians showing the different patterns of vertebral ossification in a labyrinthodont (A) and lepospondyls (B & C). The labyrinthodont is represented by *Branchiosaurus salamandroides*, from the Upper Carboniferous of Nýřany, Czech Republic. The lepospondyl is *Hyloplesion longicostatum* from the same locality, showing lateral and ventral views of vertebrae. Neural arches and centra are both well ossified in the smallest known individuals. The limb bones show the immaturity of these specimens in lacking ossification of the articulating surfaces and tarsals. The limbs and skull are at a roughly comparable stage of ossification in the labyrinthodont, but only the neural arches are beginning to ossify and remain paired. The centra show no evidence of ossification.

anthracosaurs, and like our own, it allowed movement in all directions. In contrast, the first cervical is unipartite in all lepospondyls and modern amphibians, and the articulation between the occiput and the first cervical in most of these groups restricted motion to dorsoventral flexion of the head on the trunk, with essentially no lateral bending or rotation. Although this is only one character complex, it is both complicated and has strong functional significance. A nearly identical phalangeal count of both the manus and the pes in early amniotes and anthracosaurs also supports common ancestry, distinct from that of any lepospondyls. Unfortunately, these few characters are not sufficient to balance the much larger number that are clearly associated with small size and related changes in patterns of vertebral development, trunk elongation and limb reduction.

Conclusions

In establishing relationships, it is vital to recognize that characters may be homologous on the basis of structural and positional correspondence, or similarity of developmental patterns and constraints, without necessarily being the result of immediate common ancestry.

Character transformations associated with trunk elongation and limb reduction as well as those resulting from developmental accommodation to small size may involve many traits that are highly correlated with one another, but are not necessarily the result of inheritance from a close common ancestor. In extreme cases, such as those exemplified by very small Carboniferous tetrapods, similarities due to homoplasy may far outweigh those resulting from phylogenetic homology, thus making parsimony an unreliable means of recognizing synapomorphies.

Ultimately, historical or phylogenetic homology of derived characters can only be established through increased knowledge of the fossil record, including the discovery of specimens from early members of the derived clades and their immediate common ancestors.

Acknowledgements

I thank Pamela Gaskill, who was responsible for the original specimen drawings, Lee Bar-Sagi, who assembled the data for Table 1 and Fig. 1, and Elena Roman, for scanning and arrangement of the figures and other editorial assistance. Research on which this paper was based is supported by the Natural Sciences and Engineering Research Council of Canada.

References

Bolt JR 1991 Lissamphibian origins. In: Schultze H-P, Trueb L (eds) Origins of the higher groups of tetrapods. Comstock Publishing Associates, Ithaca, p 194–222

Carroll RL 1995 Problems of the phylogenetic analysis of Paleozoic choanates. Bull Mus Natl Hist Nat Paris 4e sér 17 Section C no 1–4:389–445

Carroll RL, Chorn J 1995 Vertebral development in the oldest microsaur and the problem of 'lepospondyl' relationships. J Vert Paleon 15:37–56

Carroll RL, Bossy KA, Milner AC, Andrews SM, Wellstead C 1998 Lepospondyli. Encyclopedia of paleoherpetology, Part 1. Verlag Dr. Friedrich Pfeil, München

Clack JA 1999 The origin of tetrapods. In: Heatwole H, Carroll RL (eds) The evolutionary history of amphibians. Surrey Beatty & Sons, Chipping Norton, Australia, in press

Coates MI 1996 The Devonian tetrapod *Acanthostega gunnari* Jarvik: postcranial anatomy, basal tetrapod interrelationships and patterns of skeletal evolution. Trans R Soc Edin Earth Sci 87:363–421

Gans C 1975 Tetrapod limblessness: evolution and functional corollaries. Am Zool 15:455–467

Greer AE 1991 Limb reduction in squamates: identification of the lineages and discussion of the trends. J Herpetol 25:166–173

Hall BK, Miyake T 1995 Divide, accumulate, differentiate: cell condensation in skeletal development revisited. Internat J Dev Biol 36:881–893

Hanken J 1993 Adaptation of bone growth to miniaturization of body size. In: Hall BK (ed) Bone. Vol 7: Bone growth. CRC Press, Boca Raton, Florida, p 79–104

Heatwole H, Carroll RL 1999 The evolutionary history of amphibians. Surrey Beatty & Sons, Chipping Norton, Australia, in press

Jarvik E 1996 The Devonian tetrapod *Ichthyostega*. Fossils Strata 40:1–213

Laurin M, Reisz RR 1997 A new perspective on tetrapod phylogeny. In: Sumida SS, Martin KLM (eds) Amniote origins. Academic Press, New York, p 9–59

Lebedev OA, Coates MI 1995 The postcranial skeleton of the Devonian tetrapod *Tulerpeton curtum* Lebedev. Zool J Linn Soc 114:307–348

Lee M 1999 Convergent evolution and character correlation in burrowing squamate reptiles: a phylogenetic analysis. Biol J Linn Soc, in press

Milner AR 1993 The Paleozoic relatives of lissamphibians. Herpetol Monogr 7:8–27

Panchen A, Smithson TR 1988 The relationships of the earliest tetrapods. In: Benton MJ (ed) The phylogeny and classification of the tetrapods vol 1. Clarendon Press, Oxford (System Ass Special Vol 35A) p 1–32

Schultze H-P, Bolt JR 1996 The lungfish *Tranodis* and the tetrapod fauna from the Upper Mississippian of North America. Spec Pap Palaeontol 52:31–54

Shubin NH 1994 History, ontogeny, and evolution of the archetype. In: Hall BK (ed) Homology: the hierarchical basis of comparative biology. Academic Press, San Diego, CA, p 250–271

Wagner GP 1994 Homology and the mechanisms of development. In: Hall BK (ed) Homology: the hierarchical basis of comparative biology. Academic Press, San Diego, CA, p 273–299

Werneburg R 1989 Labyrinthodontier (Amphibia) aus dem Oberkarbon und Unterperm Mitteleuropas-Systematik, Phylogenie und Biostratigraphie. Freib Forsch-H C 436:7–57

DISCUSSION

Greene: If the phenomena you described were general, would this mean that we have little hope of ever homologizing any soft structures, physiologies, behaviours, etc.?

Carroll: Problems occur when there are extreme changes over a short period of time and when we lack knowledge of the fossil record. Phylogenetically, a similar problem exists in the interrelationship of the placental orders — we still have almost no idea of how placental orders are related to one another. On the other hand, no one questions the fact that they are monophyletic, and that the placental pattern of reproduction was homologous in this group. Similarly, there is a large amount of convergence among lizards, but we know so much about modern lizards that even without an adequate fossil record we can be certain that trunk elongation has evolved repeatedly in separate lineages. It isn't a cry of despair, but a call for constraint and being alert to the fact that homoplasy is an important factor.

Maynard Smith: You mentioned the remarks of Colin Patterson, who said 'synapomorphies are homologous', but I don't know what he means. This statement is meaningless to me.

Carroll: Any similar, derived characters that are established as synapomorphies by phylogenetic analysis are considered to be homologous, which is a kind of tautology. When this statement was published, Patterson (1982) was at the height of his insistence that the fossil record could contribute nothing to an understanding of relationships, and that the only way this could be achieved was through phylogenetic analysis, based on the most parsimonious transformation of characters. Reliance on parsimony assumes that most particular character transformations are unique and that convergence is comparatively rare. If there are situations in which homoplasy is more common than the unique origin of characters, then this won't work, but Colin Patterson is arguing that parsimony always works.

Maynard Smith: You offered as a contradiction to his position the fact that there are lots of lizards that are limb-less, and these are not necessarily closely related or descended from some common ancestors. Presumably, if you did a sensible analysis of those lizards, taking into account not only their lack of limbs but also everything else about them, 'limb-less' would not turn out to be a synapomorphy.

Carroll: Lee (1999) has recently analysed relationships among amphisbaenids, snakes and dibamids with two data sets. If the database is restricted to living forms, then limblessness is a homology that unites the three groups, but if he makes use of the fossil record, he finds that the three clades converge toward one another. Inclusion of the fossil record changes the results; in contrast, Patterson argued that living forms set the standard and the fossil record can never overturn data from living forms.

Meyer: The point is that you can only identify what a synapomorphy is if you do a global analysis on all of the characters. Some will turn out to be homoplasic and others will be synapomorphic, and you can't anticipate which will be which.

Wagner: You have brought up one of the many sources of confusion in this area. The term 'homology' has been applied to two aspects: on the one hand, to character states such as elongation, number of vertebrae and spatial relationships; and on the other hand, to developmentally individualized structures such as eyes, brains and limbs. From a biological point of view there is a vast difference between the two. It is easy to change character states once the character is established, and also to lose a character repeatedly. In such situations one would expect a high frequency of homoplasy because these characters are defined in a way that they can easily assume different character states. However, it is relatively rare to repeatedly generate an individualized character, i.e. a body part. Usually, the good indicators of large, old groups are novel structures, such as the autopodium. Unfortunately, these events are rare, and sometimes we encounter systematic problems when the innovations are not easily recognizable and identifiable, or not present at all. This way of thinking does not apply to character states, such as length and proportion, so we should try to be clear about whether we are talking about characters or character states when we are talking about homologous synapomorphy of character states.

Striedter: One of the reasons why cladistics is so appealing is that it doesn't weight characters. If you consider all the bones in a limb, they may count as many equal characters in a cladistic analysis, but loss of all these characters may be due to a relatively simple developmental modification, leading to the loss of a limb. Therefore, one must question the assumption that characters should never be weighted. But how is one to make decisions about character weights?

Wagner: No authority makes these decisions. The question is whether we are dealing with a developmentally individualized character or not. For a particular species this question can be answered experimentally. The fact that a fly eye or a limb bud, for example, can be induced elsewhere on the body, is an operational criterion for developmental individuality.

Roth: Do you see structures and character states as categorically distinct entities or do they represent two ends of a spectrum?

Wagner: They do represent two ends of the spectrum, but there is no uniform distribution of instances along the spectrum, just like in the case of the species concept. There is no sharp distinction between two conspecific populations and two populations from different species, but the majority of cases are found at either end of the spectrum. For biological concepts there are always grey zones between the extremes, but this is true in general and should not worry us here in particular.

Wilkins: If one has a lot of characters in the fossil evidence, each one of which is a bit dicey, how reliable are the cladograms at the end of the day, and how large are the data sets?

Carroll: In general, the more data you have the less well established they are.

Meyer: I agree. The consistency decreases with increasing numbers of characters and taxa.

Hall: Is it not difficult to answer a basic question like 'are amniotes monophyletic or polyphyletic?'

Carroll: For amniotes it's simple because the earliest members of each group are similar to one another. Lepospondyl orders are more difficult because each one is loaded down with autapomorphies that do not contribute to classification, and conversely there are fewer synapomorphies to support relationships.

Wray: This means that one should start off with the a priori assumption that structures are homologous until proved otherwise.

References

Lee M 1999 Convergent evolution and character correlation in burrowing squamate reptiles: a phylogenetic analysis. Zool J Linn Soc, in press

Patterson C 1982 Morphological characters and homology. In: Joysey K, Friday AE (eds) Problems of phylogenetic reconstruction. Academic Press, London (Systematics Association Special Volume Series 21) p 21–74

Generation, integration, autonomy: three steps in the evolution of homology

Gerd B. Müller and Stuart A. Newman*

*Department of Anatomy, University of Vienna, Währingerstrasse 13, A-1090 Wien, Konrad Lorenz Institute for Evolution and Cognition Research, Adolf-Lorenz-Gasse 2, A-3422 Altenberg, Austria, and *Department of Cell Biology and Anatomy, New York Medical College, Valhalla, New York 10595, USA*

Abstract. The homology concept harbours implicit assumptions about the evolution of morphological organization. Homologues are natural units in the construction of organismal body plans. Their origin and maintenance should represent a key element of a comprehensive theory of morphological evolution. Therefore, it is necessary to understand the causation of homology and to investigate the mechanisms underlying its origination. The study of this issue cannot be limited to the molecular level, because there appears to exist no strict correspondence between genetic and morphological evolution. It is argued that the establishment of homology follows three distinct (if overlapping) steps: (a) the generation of morphological building elements; (b) the integration of new elements into a body plan; and (c) the autonomization of integrated construction units as lineage-specific homologues of phenotypic evolution. In contrast with traditional views, it is proposed that the mechanistic basis for steps (a) and (b) is largely epigenetic, i.e. a consequence of the inherent propensities of developmental systems under changing conditions. Step (c) transcends the proximate mechanisms underlying the establishment of homologues and makes them independent attractors of morphological organization at the phenotypic level.

1999 Homology. Wiley, Chichester (Novartis Foundation Symposium 222) p 65–79

Homology reflects the central principle of morphological organization in organismic evolution. It is a phenotypic concept derived from macroscopic levels of observation. Initially developed as a tool in comparative anatomy and later systematics (see Panchen 1999, this volume), the homology concept is based on the recognition of identity of corresponding characters (homologues) in different species, irrespective of their variant forms or functions. This fundamental characteristic of organismic design is not addressed by current evolutionary theory. Although the theory implicitly uses homologues as quantitative characters, it regards their existence as given and concentrates on their inheritance and variation. But homologues are the *natural* units of organismal body plans, a key feature of phenotype construction, and a comprehensive theory of morphological evolution must also explain their origin and fixation. Therefore, it is necessary to understand the causation of homology and to investigate the processes underlying the evolutionary emergence and organization of homologues.

The study of this problem cannot be restricted to a molecular level of analysis. For one, there appears to exist no strict congruity between genetic and morphological evolution (see below and this volume: Raff 1999, Wray 1999). Nor is there always a close correspondence between genetic identity and morphological identity in modern organisms (Müller & Wagner 1996). And it is not evident — or even to be expected — that the proximate relationships between phenotype and genotype encountered in extant systems of development (this volume: Meyer 1999, Holland 1999) are equivalent to those that existed during their inception. Although genes play a significant role in controlling evolved development, they must not automatically be taken as the primary agents in *causing* the (homologous) configurations they help to realize today. Rather it is necessary to explore form-generating systems in their totality, which includes genetic control but also the generic material properties of, and interaction dynamics among, cells, tissues and their environment. These properties are not explicitly specified by the genome and yet they determine the realization of organismal form in a fundamental way. Collectively these factors have been termed 'epigenetic properties of development', and 'epigenetic' is used in this sense hereafter.

We will argue that the mechanistic basis for the emergence of homology in phenotypic evolution is largely epigenetic, i.e. that homology arises as a consequence of the propensities of developmental systems under Darwinian and non-Darwinian modification. This does not negate the importance of genes, but emphasizes that genetic evolution takes place in a lineage-specific epigenetic context that itself evolves and that is more determinative for biological form than genes *per se*. We propose that epigenetic factors affect morphological evolution in two important ways. One is generative — causing new structures to arise. The other is organizational — causing the integration of structures into fixed assemblies.

The generation of morphological building elements

Organisms are built of an essentially limited array of primary structures composed of simple cell arrangements and their reiterations, such as sheaths, layers, rods, tubes, and compact or hollow aggregations, often branched or segmented. Combinations of these gave rise to a number of primary body plans, to which increasingly more design is added in the course of evolution. For this reason it was recently proposed that epigenetic factors played different generative roles in two distinct phases of morphological evolution (Newman & Müller 1999). One was a 'pre-Mendelian' phase, in which the generic properties of cells and materials had a direct influence on the generation of form but were not yet made routine by heritable systems of development. The second was (and remains) the 'Mendelian' phase, in which a closer association between genome and development ensures a stabilized and heritable generation of form, but in which epigenetic properties are causal for smaller innovations and additions to the designs emerging from the first phase.

In the 'pre-Mendelian' phase both the physical properties of cells and primitive multicellular assemblages would have been the primary agents for the generation of the first organic forms leading to a range of metazoan body plans—some of which are preserved in the Vendian and Cambrian fossil faunas (Conway Morris 1993). The underlying assumption is that once cells had evolved adhesiveness, the generation of a limited number of primary forms automatically resulted, depending only on such parameters as strength of 'stickiness', cell size and cell number. Together with further refinements of cell adhesion (such as polarity of expression, and spatiotemporal modulation and qualitative diversification of adhesiveness) this accounts for virtually all design principles of simple metazoans. Lumen formation, invagination, compartmentalization, multi-layering, segmentation, etc., were all shown to be consequences of the generic physical properties of cells and condensed, chemically active tissue masses (Newman 1993, 1994, 1995, Newman & Comper 1990). These early forms would thus not require programmes for development but spontaneously 'fall into place', rather like proteins that assume tertiary structures generic to their molecular and thermodynamic properties.

We have proposed (Newman & Müller 1999) that such generic structures provide morphogenetic templates for a biochemical sophistication of cell interaction and thus become co-opted by the genome, ensuring heritability and a more deterministic relationship between genetic circuitry and form generation (see below). But even in the subsequent Mendelian phase of organismal evolution, characterized by evolved forms of genetically stabilized development, new morphological detail was added largely as a consequence of epigenetic conditions (Müller 1990, Müller & Wagner 1991).

'Epigenetics' in this second phase includes not only generic material properties, which are still effective, but also the hierarchical organization of developmental interactions and the dynamics of inductive networks. Through the multitude of interactive processes development assumes homeostatic, systemic qualities (Bertalanffy 1932, Needham 1936, Waddington 1957). As a consequence, any Darwinian modification of a part of the system, such as a selection for changes in body proportions, interferes with epigenetic equilibria and can eventually reach critical thresholds for a particular parameter. This could be the cell number required for a blastema, the distance between inductive tissues, the concentration of a morphogen controlling pattern formation and many more. At such a threshold point the phenotypic outcome of a parameter change will depend on the developmental reaction norm of the affected system, which may result in the loss of a structure or the appearance of a new one. The primary evolutionary modification, e.g. the simple selection for the length of a body part, could thus have quite secondary consequences, resulting in morphological innovations that are a by-product of the developmental system affected. This mechanism for the origin of morphological novelties has been conceptualized as the 'side-effect hypothesis' (Müller 1990). It explains how the continuous variation of a developmental parameter can lead to discontinuous phenotypic results. According to this concept, the origin of new morphological characters, which eventually may become homologues of a body plan, would be contingent on the epigenetic structure of the developmental system acquired by an evolving lineage.

The integration of new elements into a body plan

The previous section described how generic, epigenetic properties of development suffice to generate new structures. Since the very basic generic properties are pervasive they should tend to give rise to recurrent, similar innovations in independent phylogenetic lineages. Indeed they do, as evidenced by the frequent homoplasies that occur in organismal evolution (Sanderson & Hufford 1996). Homoplasies are a rather direct result of the epigenetic principle described. The question then is: how do new characters become homologues, i.e. fixated units of established body plans, characteristic of a phylogenetic lineage? We propose that the answer lies in the increasing integration of different levels of organismal evolution, in which epigenetic factors again play a central role. The levels concerned are phenotypic, developmental and genetic.

One aspect of integration lies in the evolutionary increase of conditional developmental interactions and the creation of epigenetic and genetic networks (Abouheif 1999, this volume). Phenotypic outcomes initially brought about by generic means will come to rely on sophisticated molecular mediation that depends on genetic coding. The genome itself also evolves, significantly by the duplication of developmental control genes (Holland 1999, this volume), and a redundant genetic repertoire (Cooke et al 1997, Nowak et al 1997, Picket & Meeks-Wagner 1995, Tautz 1992, Wagner 1996, Wilkins 1997) becomes available for the control of new processes, which thereby become overdetermined (Newman 1994). The epigenetic context in which such genetic redundancy prevails will strongly determine which developmental interactions will be genetically integrated (Müller & Wagner 1996), resulting in increasingly closer mapping between genotype and phenotype as seen in most extant organisms (Newman & Müller 1999). Note that the epigenetic context in which genetic evolution takes place is itself subject to change.

A second, equally important aspect of integration lies in the structural and functional interrelations that evolve at the phenotype level. Morphological innovations, although initially a result of epigenetic conditions, are tested at the phenotype level and will be maintained and modified by selection. Numerous structure–function interrelationships will evolve (Galis 1996) and integrate new characters at the phenotypic level. Over time, the number of structural and functional interdependencies and the position of a new character in the hierarchical design of an organism will gain importance as more design and functional differentiation is added. This, again, has repercussions at the genetic level, with selection favouring the genetic linkage of functionally coupled characters (Bürger 1986, Wagner 1984).

By these means, genes, development and phenotypic structure become increasingly interdependent, resulting in the evolution of character complexes typical of organismal body plans. Epigenetic mechanisms of development are central in this process of organization of body design because they provide templates for both phenotypic and genetic integration. The sum of these processes 'locks in' new characters that arose as a consequence of the conditional mechanisms discussed above and thus generates the heritable, inter-taxonomic building units of the phenotype described as homologues. In principle, the number of developmental, structural and functional interdependencies accumulated by a homologue can be quantified. This provides a means of defining the strength of fixation of a homologue and of the probability of it becoming undone during subsequent steps of evolution, a notion embodied in the concepts of 'burden' (Riedl 1978) and 'constraint' (Maynard Smith et al 1985). The latter offers possibilities for experimental testing (Alberch & Gale 1985, Streicher & Müller 1992; see also Wagner 1999, this volume).

It is central to the concept presented here that the fixation and heritability of new elements in phenotype evolution is secondary to their origination, a view anticipated in many different forms, such as in the concepts of assimilation (Waddington 1957), generative entrenchment (Wimsatt 1986), epigenetic traps (Wagner 1989) or genomic co-optation (Newman & Müller 1999).

The autonomization of homology

In the two preceding sections we have outlined a scenario that describes how processes of generation and fixation lead to the homologues of phenotype construction. But to stop here would mean to neglect what is probably the most salient attribute of homology. There is a third and most fascinating step in the evolution of homology, characterized by an increasing independence of the constructional entities established: homology may persevere while the underlying developmental, molecular and genetic constituents drift.

Although in modern organisms there is a remarkably conserved relationship between genes and structure (see this volume: Wray 1999, Meyer 1999), empirical evidence continues to highlight the lack of concordance among genetic, developmental and morphological evolution. It is frequently observed that the same phenotypic endpoint can be reached via different developmental pathways, and that indeed such pathways have changed importantly in the case of structures that are clearly homologous (see this volume: Raff 1999, Wray 1999). Genetic control can change, inductive interactions can change, the molecular makeup can change and yet a homologue is maintained at the phenotype level (Bolker & Raff 1996, Hall 1994, Wagner 1995, Wray & Raff 1991). Experimental studies equally demonstrate that genetic and morphological variation can be poorly correlated (Atchley et al 1988). These observations of comparative developmental biology are paralleled by population genetic findings of an incongruence between rates of genetic and morphological divergence (Bruna et al 1996, Meyer et al 1990, Sturmbauer & Meyer 1992) and are further supported by the observation that genome complexity is poorly correlated with differences of form and function (Miklos & Rubin 1996).

These findings of a decoupling between genetic and morphological evolution indicate that homologues, as design units of the phenotype, eventually transcend their underlying molecular, epigenetic and genetic constituents, and begin to play an independent organizational role in morphological evolution. When integrated as indispensable parts of a bauplan, homologues assume constructional identity and come to act as matrices of morphological organization. They become 'attractors' of morphological design, around which more design is added, and will be conserved even in the face of changing adaptive conditions. Thus, homologues, as individualized building units of the phenotype, become more influential for the further path of morphological evolution than the primary generic conditions underlying their origin and the biochemical circuitry that controls their developmental establishment. Homology, at this stage, is the product of an increasing autonomization of a design principle from its mechanistic underpinnings.

Conclusion

A causal approach to homology indicates that, conceptually, three steps (generation, fixation, autonomization) must be distinguished in its establishment, although it is evident that these will overlap in reality. It is proposed that the generation of new elements is largely a consequence of epigenetic principles, while their fixation is a function of progressive integration of the phenotypic, developmental and genetic levels of evolution, with 'epigenetic templates' at their core. In a final step homologues gain autonomy as organizers of phenotype construction. Such a view has a number of theoretical consequences:

(1) The generation of organismic form—and of homology as its constituent design principle—would be based on different mechanisms than the small, gradual, adaptive variations of characters that are the objects of neo-Darwinian theory.
(2) Homologues have their origin in epigenetic properties and mechanisms that provide templates for both genetic and phenotypic organization, and are thus secondarily co-opted by genetic evolution rather than resulting from it.
(3) Theories of organismal evolution must take account of the fact that homologues can act as attractors of design in their own right, and thus become autonomous organizers of the phenotype in an evolutionary lineage.

This conception for the origin of morphological homology represents a step away from the gene-centred view of organisms, which has unduly assumed paradigmatic status in biology (Strohman 1997). It agrees with calls for a new conceptualization of the relationship between genes and form (Neumann-Held 1998, Newman & Müller 1999, Nijhout 1990, Oyama 1985), and it represents an alternative mechanistic principle underlying morphological evolution. It instils a structuralist component into evolutionary theory by including a principle of organization that results from the possible sets of transformations of form-generating systems, while integrating the results of modern evolutionary and developmental genetics. Such a conceptual move could be part of an anticipated paradigm shift in biology (Strohman 1997), leading to a more comprehensive evolutionary theory of phenotype organization, with homology as one of its central elements.

Acknowledgements

We thank Bernhard Strauss and Diego Rasskin-Gutman for valuable comments and inspiring discussions during the preparation of this manuscript.

References

Abouheif E 1999 Establishing criteria for regulatory gene networks: prospects and challenges. In: Homology. Wiley, Chichester (Novartis Found Symp 222) p 207–225
Alberch P, Gale EA 1985 A developmental analysis of an evolutionary trend: digital reduction in amphibians. Evolution 39:8–23

Atchley WR, Newman S, Cowley DE 1988 Genetic divergence in mandible form in relation to molecular divergence in inbred mouse strains. Genetics 120:239–253
Bertalanffy Lv 1932 Theoretische Biologie I. Borntraeger, Berlin
Bolker JA, Raff RA 1996 Developmental genetics and traditional homology. BioEssays 18:489–494
Bruna EM, Fisher RN, Case TJ 1996 Morphological and genetic evolution appear decoupled in Pacific skinks (Squamata, Scincidae, Emoia). Proc R Soc Lond B Biol Sci 263:681–688
Bürger R 1986 Constraints for the evolution of functionally coupled characters: a nonlinear analysis of a phenotypic model. Evolution 40:182–193
Conway Morris S 1993 The fossil record and the early evolution of the Metazoa. Nature 361:219–225
Cooke J, Nowak MA, Boerlijst MC, Maynard Smith J 1997 Evolutionary origins and maintenance of redundant gene expression during metazoan development. Trends Genet 13:360–364
Galis F 1996 The application of functional morphology to evolutionary studies. Trends Ecol Evol 11:124–129
Hall BK (ed) 1994 Homology: the hierarchical basis of comparative biology. Academic Press, San Diego, CA
Holland PWH 1999 How gene duplication affects homology. In: Homology. Wiley, Chichester (Novartis Found Symp 222) p 226–242
Maynard Smith J, Burian R, Kauffman S et al 1985 Developmental constraints and evolution. Q Rev Biol 60:265–287
Meyer A 1999 Homology of developmental genes. In: Homology. Wiley, Chichester (Novartis Found Symp 222) p 141–157
Meyer A, Kocher TD, Basasibwaki P, Wilson AC 1990 Monophyletic origin of Lake Victoria cichlid fishes suggested by mitochondrial DNA sequences. Nature 347:550–553
Miklos GL, Rubin GM 1996 The role of the Genome Project in determining gene function: insights from model organisms. Cell 86:521–529
Müller GB 1990 Developmental mechanisms at the origin of morphological novelty: a side-effect hypothesis. In: Nitecki MH (ed) Evolutionary innovations. University of Chicago Press, Chicago, IL, p 99–130
Müller GB, Wagner GP 1991 Novelty in evolution: restructuring the concept. Annu Rev Ecol Syst 22:229–256
Müller GB, Wagner GP 1996 Homology, *Hox* genes, and developmental integration. Am Zool 36:4–13
Needham J 1936 Order and life. MIT Press, Cambridge, MA
Neumann-Held EM 1998 The gene is dead — long live the gene: conceptualizing genes the constructionist way. In: Koslowsky P (ed) Sociobiology and bioeconomics: the theory of evolution in biological and economic theory. Springer-Verlag, Berlin, p 105–137
Newman SA 1993 Is segmentation generic? BioEssays 15:277–283
Newman SA 1994 Generic physical mechanisms of tissue morphogenesis: a common basis for development and evolution. J Evol Biol 7:467–488
Newman SA 1995 Interplay of genetics and physical processes of tissue morphogenesis in development and evolution: the biological fifth dimension. In: Beysens D, Forgacs G, Gaill F (eds) Interplay of genetic and physical processes in the development of biological form. World Scientific Publishing, Singapore, p 3–12
Newman SA, Comper WD 1990 'Generic' physical mechanisms of morphogenesis and pattern formation. Development 110:1–18
Newman SA, Müller GB 1999 Epigenetic mechanisms of character origination. In: Wagner GP (ed) The character concept in evolutionary biology. Academic Press, San Diego, CA, in press
Nijhout HF 1990 Metaphors and the roles of genes in development. BioEssays 12:441–446

Nowak MA, Boerlijst MC, Cooke J, Maynard Smith J 1997 Evolution of genetic redundancy. Nature 388:167–171
Oyama S 1985 The ontogeny of information. Cambridge University Press, Cambridge
Panchen AL 1999 Homology — history of a concept. In: Homology. Wiley, Chichester (Novartis Found Symp 222) p 5–23
Pickett BF, Meeks-Wagner DR 1995 Seeing double: appreciating genetic redundancy. Plant Cell 7:1347–1356
Raff RA 1999 Larval homologies and radical evolutionary changes in early development. In: Homology. Wiley, Chichester (Novartis Found Symp 222) p 110–124
Riedl R 1978 Order in living organisms. Wiley, Chichester
Sanderson MJ, Hufford L (eds) 1996 Homoplasy: the recurrence of similarity in evolution. Academic Press, San Diego, CA
Streicher J, Müller GB 1992 Natural and experimental reduction of the avian fibula: developmental thresholds and evolutionary constraint. J Morphol 214:269–285
Strohman RC 1997 The coming Kuhnian revolution in biology. Nature Biotechnol 15:194–199
Sturmbauer C, Meyer A 1992 Genetic divergence, speciation, and morphological stasis in a lineage of African cichlid fishes. Nature 358:578–581
Tautz D 1992 Redundancies, development and the flow of information. BioEssays 14:263–266
Waddington CH 1957 The strategy of the genes. Allen and Unwin, London
Wagner A 1996 Genetic redundancy caused by gene duplications and its evolution in networks of transcriptional regulators. Biol Cybern 74:557–567
Wagner GP 1984 Coevolution of functionally constrained characters: prerequisites for adaptive versatility. BioSystems 17:51–55
Wagner GP 1989 The origin of morphological characters and the biological basis of homology. Evolution 43:1157–1171
Wagner GP 1995 The biological role of homologues: a building block hypothesis. Neues Jahrb Geol Palaeontol Abh 195:279–288
Wagner GP 1999 A research programme for testing the biological homology concept. In: Homology. Wiley, Chichester (Novartis Found Symp 222) p 125–140
Wilkins AS 1997 Canalization: a molecular genetic perspective. BioEssays 19:257–262
Wimsatt WC 1986 Developmental constraints, generative entrenchment, and the innate-acquired distinction. In: Bechtel W (ed) Integrating scientific disciplines. Nijhoff, Dordrecht, p 185–208
Wray GA 1999 Evolutionary dissociations between homologous genes and homologous structures. In: Homology. Wiley, Chichester (Novartis Found Symp 222) p 189–206
Wray GA, Raff RA 1991 The evolution of developmental strategy in marine invertebrates. Trends Ecol Evol 6:45–50

DISCUSSION

Wagner: You used the phrase 'attractors' of morphological design in two different senses: (1) the attractor becomes a part of the body plan and 'attracts' innovations; and (2) the final ontogenetic state of the character becomes independent of the pathway through which the state is reached, because it is an attractor of ontogenetic dynamics.

Müller: No, with the term 'attractor' I specifically refer to the first process you mentioned, but it would represent a necessary precondition that leads to the second: developmental independence.

Wagner: I am not sure this is logically necessary. I understood from what you said that the end product becomes independent of the developmental pathway because of self-stabilizing mechanisms. This doesn't mean that this self-stabilization of the end product is not mechanistically explained, but rather cell–cell and tissue–tissue interactions make the end product a morphologically stable entity (Wagner & Misof 1993). The developmental pathway becomes more flexible and the end product becomes invariant to this pathway.

Müller: The attractors I am referring to act at the structural, phenotypic level of evolution. They are responsible for the actual configuration of the end product (the specific set of homologues that make up the bauplan of a lineage) which becomes stabilized by cell and tissue interactions. But the proximate mechanisms that stabilize the end product in development must not be regarded as the causal agents for the anatomical configurations they stabilize. Developmental self-stabilization is not the essence of the attractor part of the hypothesis.

Wagner: If the end product is self-stabilized, this stage then becomes a dynamic attractor, and therefore it is independent of the developmental sequence.

Müller: I agree that this is a possibility, but the maintenance of the attractors would be a consequence of the organizational role they come to play at the constructional and functional levels of organismal evolution.

Rudolf Raff: Could you tell me whether the muscle attachment on the fibula opposite the fibular crest of the tibiotarsus in birds is present in crocodiles or other reptiles?

Müller: Yes, it is.

Rudolf Raff: So, why don't they form a crest?

Müller: Because the biomechanical epigenetic conditions are different in avian limb development as compared with reptilian limb development. As a result of a progressive evolutionary reduction of the fibula, birds have a short and thin cartilaginous rudiment of the fibula, and the pulling action during active movement of the embryo, transmitted from the iliofibularis muscle inserting on the fibula, increases the mechanical load on the connective tissue between the fibula and the tibiotarsus. The ability of connective tissue cells to respond to pressure stimuli with cartilage matrix expression causes the formation of the cartilaginous sesamoid on which the final osseous crest depends (Müller & Streicher 1989). Reptiles do not have a crest because their fibula is long and thick, except for theropod dinosaurs in which the fibula is also strongly reduced—presumably also in their embryos. Selection is involved in the process of fibula reduction, but the specificity of the morphological outcome of this process (the fibular crest) is provided by the developmental system.

Akam: But your own data suggest that on top of this there is a process which causes the specificity to become hardwired. All of the lumps and bumps of the basic bone structure are independent of use and the different crests, or different additional cartilages, are to a differential extent already becoming independent of movement.

Müller: Yes, this is correct, as you say 'to a differential extent'. But our data also show that as long as the epigenetic conditions remain the same, there is little requirement for genetic hardwiring.

Wilkins: I like the emphasis in your work on biophysics in development, but it is not possible to convert this directly into evolutionary changes without involving genetic changes that do the same thing. Eventually, gene products have to be made that mimic the physical conditions creating the developmental change.

Müller: This is a possibility. I have no problems with that. My hypothesis concerns the origin of new structures, and not so much the mechanisms of their further maintenance and variation, which I believe are quite Darwinian.

Hall: It is the phenotypic plasticity that is heritable, and within that there is a range over which morphological change can occur.

Wilkins: I would say that what you're proposing is close to the genetic assimilation hypothesis of Waddington (1961).

Roth: But genetic assimilation as a model for how a trait comes to be reliably expressed is well understood. A trait is what is selected. Any alleles that more consistently evoke it will be favoured, and this can involve any number or combination of alternatives.

Müller: There must exist causal mechanisms for the origin of new characters in the first place, before assimilation can take place. I do not deny that innovations can also occur through genetic change, but there are few cases known where mutations of structural genes cause morphological innovation, although bird feathers are probably a good example, as shown by Brush (1993). And it has not been demonstrated that duplications of regulatory genes were directly linked to morphological changes in phenotypic evolution. I believe that such duplications generate a redundancy of regulatory modules which can later be used to control new developmental interactions that stabilize new processes and the characters that result from them.

Nicholas Holland: We don't yet know much about how cell-specific changes translate into tissue-specific changes in development. You may be right at the end of the day, but for now this is a weak point in your argument. There are few examples of developmental genes directly upstream of genes involved in controlling tissue architecture via changes in the extracellular matrix, although one is that the *pax8* product directly affects the gene encoding N-CAM, according to Edelman & Jones (1995).

Müller: I agree, but by the term 'pre-Mendelian' I was referring to a pre-multicellular and early multicellular world—there would not have been much around yet in the way of tissues.

Tautz: I have a problem with the use of the term epigenetics. In my view it means we don't understand the genetics, so we call it epigenetics.

Müller: I don't agree with this, but it may help to point out that there are two different uses of the term epigenetics. It is important to distinguish between epigenetic inheritance and epigenetic development. These are two separate problems, although they may be related, which could be the most interesting part. Here I am speaking of epigenetics in development, which, in my view, is the sum of all factors that have a causal influence on gene expression or other processes of development in determining the morphological outcome. This includes physicochemical factors, geometrical factors, biomechanical factors, generic properties of cells and tissues, and the dynamics of developmental interactions. Pre-Mendelian epigenetics is primarily a function of the physicochemical properties of condensed materials, to which belong primitive adhesive cell masses. The cells involved have no developmental programming that leads to an organized form, rather they divide and associate repeatedly, and automatically produce the basic, generic forms I have shown, such as spheres, or tubes, or layers, etc. Mendelian epigenetics on the other hand, while also dependent on physicochemical properties, is primarily a function of the interaction dynamics in heritable developmental systems.

Tautz: Why should a cell associate with another cell if it has a better chance of surviving on its own?

Müller: That is a problem of the origin of multicellularity. We know, at some point cells did stick together! This adhesive property may well have been a consequence of cell surface proteins that in turn are gene products. But this is not the point. The point I am trying to make is that the morphological forms that result from the association of a number of adhesive cells are a consequence of the dynamic systems properties of such multicellular aggregates, and these, in a pre-Mendelian world, would strongly depend on physicochemical conditions.

Tautz: But that step would have been advantageous. It would not have been a chance effect.

Lacalli: I don't think it was a chance effect, nor that it happened quickly. Consider the elegantly simple multicellular stages of early embryogenesis: the blastula and gastrula. It is possible that such simplicity evolved quickly with a few mutations. I suspect the opposite, however, that the apparent simplicity of these kinds of cell assemblages is really the result of a long sequence of complicated genetic changes. Similarly, the striped *ftz* pattern in *Drosophila*, although it looks simple, is generated at the molecular level in a complicated way.

Maynard Smith: I wonder whether we actually all agree in a curious way. No geneticist imagines that genes programme embryos without the laws of physics and chemistry. Processes such as cell movement, adhesion, folding and contraction generate shapes. Gene changes ultimately can only alter morphology by altering the behaviour of dynamic systems, even if these dynamic systems are extremely simple, such as a dynamic system that folds a protein. Proteins don't fold themselves up, they fold up because of the laws of physics and chemistry. The point of disagreement is purely in the semantics of choosing a new language for these processes. Gerd Müller is not saying anything substantially different from what I've always thought.

Wagner: We all agree that physics and chemistry are involved, but there might be a disagreement about what is the most informative level to look at in terms of experimentation if we want to understand the origin of a new body part. I would say that Gerd left out one mode of innovation likely to be primarily related to genetic changes, i.e. the differentiation of repeated elements, such as segments; but this is not the only mode of innovation. Gerd gave a number of other examples.

Müller: According to our own definition (Müller & Wagner 1991) serial characters and their differentiations do not constitute true 'innovations'—this is why I did not mention them in the present context.

Tautz: It's far too early to speculate about other factors, because we don't even understand the genetic factors that influence morphology. There are no published examples of regulatory changes that have caused an innovation. However, we have published a paper in which we traced the ancestral regulatory changes that are involved in the transition from short-term to long-term embryogenesis in insects (Wolff et al 1998). Once we understand the genetics we may not need to invoke any other factors.

Maynard Smith: You were asking earlier why a cell should associate with another cell. Presumably, you would agree that (a) it becomes sticky because some gene makes a protein that makes the cell sticky, and (b) that the population becomes sticky because it is an advantage to be sticky. But it does this through a physical process, and Gerd Müller is not saying anything other than that. He is saying it is a physical process and you are saying it's a genetical one, but why do we have to choose between the two?

Müller: We do not have to choose at all. It would be sufficient to agree on the fact that genes are not the only causal agents responsible for the evolutionary origin of the forms whose establishment in ontogeny they help to control today, and that a comprehensive theory of morphological evolution will have to take account of a number of epigenetic causalities, of which the physical ones represent but one category. Diethard Tautz's comments indicate that we may be far from such a consensus.

Roth: We don't know enough about the genetic basis of these processes, so that's one direction of interest. Another direction of interest, however, deals with the physical laws that effect reorganization at the morphological level, and these may be screened off from the changes in gene expression. We are talking about two different research programmes, and although they interdigitate — they have causal connections — they employ different techniques and approaches.

Hall: Presumably, once we know all the genes involved in generating a particular phenotype, we would still agree that genes do not make phenotypes, and that there is an intermediate level between the gene and the phenotype.

Tautz: I don't see why, once we have identified the genes and their properties, and particularly those that are responsible for creating cell shape, we should be able to understand the phenotype. The properties they have are what I would call epigenetics.

Hinchliffe: This emphasis on epigenetic factors reminds me of an experiment we did on 'extradigit' formation in the limb buds of chick embryos (Hinchliffe & Horder 1993). We took a relatively well-developed embryonic leg in which digits were already forming and made an incision in the mesenchyme tissue between posterior digits 3 and 4. We found that the interdigital mesenchyme condensed and then underwent an entire process of morphogenesis to form a digit with sometimes 2 or 3 phalangeal elements. What's interesting about this is that tendons formed in association with the extra digits. This process of repatterning (the normal fate of this mesenchyme is cell death) occurs relatively late compared with the normal digit formation process controlled by the zone of polarizing activity. We don't yet know the particular genes that confer identity normally on specific digits (the usual suspects are expressed too early to explain the late-forming extra digits), or whether indeed this is the right sort of approach. However, this experiment does show that the relationship between pattern-regulating genes and the morphological outcome is sometimes indirect.

Wagner: I recently looked into the question of whether all phalanges form as a result of pattern formation or whether additional phalanges form after the digits individualize. What generally happens is that after digit individualization, not only do the digits grow, but new phalanges are also added. Therefore, the preliminary data suggest that in all tetrapods there is a phase in which new skeletal elements are added after the pattern formation events cease. This is interesting because the genes involved in pattern formation seem to be redeployed at this later stage.

Akam: This idea of development after pattern formation seems bizarre to me. Pattern formation is a continuously iterative process, such that each pattern is the starting condition for the next process of pattern formation. It doesn't make any sense to say that one process is pattern formation and the later one is not.

References

Brush AH 1993 The origin of feathers: a novel approach. In: Farner DS, King JR, Parkes KC (eds) Avian biology, vol 9. Academic Press, London, p 121–262

Edelman GM, Jones FS 1995 Developmental control of N-CAM expression by *Hox* and *Pax* gene products. Philos Trans R Soc Lond B Biol Sci 349:305–312

Hinchliffe JR, Horder TJ 1993 Lessons from extradigits. In: Fallon JF, Goetinck PF, Kelley RO, Stocum DL (ed) Limb development and regeneration, vol A. Wiley-Liss, New York, p 113–126

Müller GB, Streicher J 1989 Ontogeny of the syndesmosis tibiofibularis and the evolution of the bird hindlimb: a caenogenetic feature triggers phenotypic novelty. Anat Embryol (Berl) 179:327–339

Müller GB, Wagner GP 1991 Novelty in evolution: restructuring the concept. Annu Rev Ecol Syst 22:229–256

Waddington CH 1961 Genetic assimilation. Adv Genet 10:257–293

Wagner GP, Misof BY 1993 How can a character be developmentally constrained despite variation in developmental pathways? J Evol Biol 6:449–455

Wolff C, Schröder R, Schulz C, Tautz D, Klingler M 1998 Regulation of the *Tribolium* homologues of *caudal* and *hunchback* in *Drosophila*: evidence for maternal gradient systems in a short germ embryo. Development 125:3645–3654

On the homology of structures and *Hox* genes: the vertebral column

Frietson Galis

Institute for Evolutionary and Ecological Sciences, University of Leiden, PO Box 9516, 2300 RA Leiden, The Netherlands

Abstract. Research on expression patterns of *Hox* genes has revealed a surprisingly high conservation among vertebrates. In agreement with this conservation, a correlation has been found between the anterior limits of expression areas of certain *Hox* genes and the borders between morphological regions of the vertebral axis. These similarities are striking and important, but also counterintuitive, unless there are strong selection pressures to protect this conservatism. It is important to identify the selective forces that maintain these conservative networks. These selective forces can be due to pleiotropy or to internal selection. Discussed are the selective factors that are involved in the evolutionary constraint on the number of cervical vertebral numbers in mammals. Factors involved are due to internal selection and involve susceptibility to cancer, stillbirths and neuronal problems. It is intriguing how similar genetic networks can lead to fundamentally different animals. Clearly the same genes are used for different purposes. It is therefore important to try to find these differences. The search for homology between organisms, and the enthusiasm about similarities that come with it, at times impedes the discovery of such differences. I have searched the literature for differences within vertebrates in the functioning and expression patterns of *Hox* genes during the development of the vertebral axis. The ensuing implications for homology of structures and genes are discussed. The vertebral column is a promising model system for the evaluation of the relationship between homologous *Hox* genes and homologous structures because of the large conservation of *Hox* gene expression patterns along the anterior axis. However, extensive remodelling of the vertebrate column indicates that important changes in the genetic basis must have taken place. A survey of the literature indicates that the correlation between *Hox* gene expression areas and vertebral regions is not such that one can predict the borders between vertebral regions on the basis of *Hox* gene expression patterns. The involvement of *Hox* genes in the development of identity of vertebrae is complex and the problems regarding the value of gene expression patterns for the determination of anatomical homology are discussed.

1999 Homology. Wiley, Chichester (Novartis Foundation Symposium 222) p 80–94

The study of the expression patterns of *Hox* genes and other homeobox genes has revealed a surprisingly high conservation among vertebrates, and even between vertebrates and insects conservation is high. The functions of these genes show surprising conservation as well, although the differentiation of functions during evolution is as impressive as its conservation. The constancy in expression and functioning is counterintuitive, as the selective forces on these organisms vary greatly and have resulted during evolution in significant structural and functional differences at a phenotypic level. There are thus two important issues to understand: (1) which differences in the genetic organization are responsible for the striking organismal variety; and (2) what are the powerful selective factors protecting the conservation?

The conservation of the homeobox genes provides a unique opportunity for comparative research and thus for the study of homology. Indeed, similarities in gene expression can throw light on the evolutionary history of seemingly unrelated structures and processes. On the other hand, the acquisition of new functions and the loss of old functions for homeobox genes present a real danger because they will lead to the involvement of genes with different structures and tissues. The many homeotic transformations found in transgenic mice with mutations in *Hox* genes suggest that the relationship between genes and structures is better understood than is actually the case. To better understand the gap between gene expression and structures, which is necessary to evaluate the agreement between homology of *Hox* genes and structures, these two questions need to be answered.

Assessment of differentiation and conservation

Differences in Hox gene expression between species belonging to different vertebrate classes

The coincidence in expression patterns of *Hox* genes between *Drosophila* and mice during development is the most striking example of widespread conservation of developmental genes in animals. It is thus surprising that the expression patterns of *Hox* genes in the teleostean zebrafish turns out to be considerably different from both mice and *Drosophila* (Prince et al 1998a). In zebrafish all *Hox* genes have rostral borders that lie close together in the rostral part of the body, both in the CNS and in the paraxial mesoderm. Importantly, the relative order of the rostral borders of *Hox* genes (spatial colinearity[1]) is preserved despite the compression of rostral

[1]In vertebrates there are four duplicated clusters of *Hox* genes—*HoxA*, *HoxB*, *HoxC* and *HoxD*—located on separate chromosomes (e.g. Krumlauf 1992). Each cluster contains 13 genes in an ordered sequence, or less when genes were lost during evolution. The ordering of the sequence is, with few exceptions, the same for the physical location of the gene on the chromosomes as for the expression pattern within the developing embryo along the anteroposterior axis. This phenomenon is called spatial colinearity. Paralogous genes are genes that are presumably derived from the same gene before the duplication of the clusters, thus *Hoxa4*, *b4*, *c4* and *d4* are paralogues of group 4).

borders within a small area. Prince et al (1998a) suggest that the pattern of mice is derived and has evolved from a zebrafish-like situation in response to selection for more complexity in parts caudal to the head. In the light of the stunning conservation of *Drosophila* and mice, it is more plausible that the situation in zebrafish is derived. A derived state in zebrafish is not surprising given the long independent evolutionary development of actinopterygian fish since the Silurian (tetrapods have evolved from crossopterygian fish). It is likely that the abundance of *Hox* gene expression in the head region is causally related to the complexity of the head of teleost fishes. I propose, therefore, that the rostral shift of expression patterns occurred in response to selection for more differentiation of structure and function in the head of teleost fish. The structural complexity of the head of teleosts has played an important role in the evolution of this clade (Janvier 1996) and is presumably a causal factor for its unequalled species richness among vertebrates (Galis & Metz 1998).

Importantly, these data show that deviations in the expression patterns from the primitive pattern do occur and are possibly more widespread than was assumed until recently. There are more differences in *Hox* genes between fish and mammals. For instance, the group 4 and 5 *Hox* genes have a low to negligible expression in the paraxial mesoderm in zebrafish except for *Hoxb4*, whereas all genes of these groups are expressed in mice in the paraxial mesoderm (Prince et al 1998a). In *Fugu rubripes,* another teleost fish species and one characterized by a highly specialized morphology (absence of ribs), more differences regarding *Hox* genes were found, i.e. a complete lack of six or seven genes, most notably of the D cluster (Aparicio et al 1997).

In *Xenopus,* unlike in zebrafish and mice, *Hoxb* genes are not expressed in somites and their derivatives except for *Hoxb4* and *Hoxb5* (Godsave et al 1994). In addition, the expression patterns of *Hoxb* genes are different in the CNS . *Hoxb4* has again an exceptional expression pattern in that it is restricted to the future hindbrain. *Hoxb5*, *Hoxb7* and *Hoxb9* have the same expression pattern and are expressed throughout the spinal cord. There is thus a breakdown of spatial colinearity in the CNS for these *Hox* genes (a similar breakdown of spatial colinearity is found in zebrafish for *Hoxb7*, *Hoxb8* and *Hoxb9* because they have the same rostral border both in the CNS and the paraxial mesoderm, Prince et al 1998a).

When interpreting data of larval stages of *Xenopus* it should be kept in mind that in *Xenopus,* like in other anurans, the development of the larvae is strongly decoupled from the development of adult structures during metamorphosis. Many adult structures are not formed from larval cells, but from set-aside cells. One therefore expects important differences in the genetic organization of development.

In birds differentiation of function of *Hox* genes along the anteroposterior axis has occurred as well. In quails *Hoxa7* is, unlike other *Hox* genes, not involved in the pattering of the axial skeleton, but has been co-opted for a different function (Xue et al 1993). After the division of the somites into a sclerotomal and a dermomyotomal part, expression of *Hoxa7* becomes restricted to the myotome of all somites and is thus involved in the development of trunk and limb musculature (Gossler & Hřabe de Angelis 1998; the axial skeleton is derived from the sclerotomal part of somites). This expression pattern differs from that in zebrafish, mice and humans where the expression is not in all somites and, furthermore, is restricted to the sclerotomal part of the somites (Prince et al 1998a, Knittel et al 1995). The expression and functioning of this gene in the CNS is also aberrant. It is the only *Hox* gene of Antp type which is not only expressed in the hindbrain, but is expressed throughout the CNS, including the forebrain and eyes. In the CNS it has a function in the proliferation of cells (Xue et al 1998). Interestingly, in *Xenopus,* another *Hox* gene, *Hoxc6*, appears to be exceptional in not being involved in the patterning of the axial skeleton, but in the development of myotomal structures and the nervous system (Wright et al 1989).

These differences between vertebrate species of difference classes present evidence that evolution does model the functions and expression patterns of *Hox* genes even when expressed along the anteroposterior axis.

Differentiation of spatial colinearity within organisms

To a certain extent the expression of *Hox* genes in many tissues and structures bears a relation with the order of the genes in the clusters on the chromosomes. However, there is good agreement only along the anteroposterior axis in the CNS and the paraxial mesoderm. In all other cases, important deviations from spatial colinearity have evolved in parallel with a differentiation of function. For example, in the limbs differential expression patterns of *Hox* genes have evolved for the specification of identity (i.e. fore- versus hindlimb, proximal versus distal part of limbs) in a way that is not well correlated with the order of the genes along the chromosomes (Rijli & Chambon 1997). Other *Hox* genes have apparently lost their regional restriction and are expressed throughout the limb, e.g. *Hoxa11* has an almost global expression in the chick limb mesoderm and seems to be involved in the repression of chondrogenic differentiation (Rogina et al 1992). Similarly, in the skin some *Hox* genes are globally expressed whereas others are regionally restricted (Godwin & Capecchi 1998). In blood, the expression of specific *Hox* clusters appears to be restricted to specific cell lineages. Thus, *Hox* genes of the A cluster are found in myelomonocytic cells, of the B cluster in erythropoietic cells and of the C cluster in lymphopoietic cells (Zimmermann & Rich 1997). Furthermore, there are comparable differences in timing, e.g. *Hoxb7* is expressed at all times whereas *Hoxb8* expression is restricted in time (7.5–11.5 days post coitum).

Even the expression patterns that show the most conserved spatial colinearity, those in the paraxial mesoderm and CNS (but see above for *Hoxa7* in quails), do differ considerably from each other. In general *Hox* expression in the neural tube is offset rostrally by several somites relative to the expression in the paraxial mesoderm (e.g. for *Hoxb4* six somites in mice and one somite in zebrafish; Lufkin 1996, Prince et al 1998a). Thus, the precise molecular mechanisms which direct *Hox* gene expression at a given axial level are probably different between the neural tube and adjacent mesoderm (Lufkin 1996) and this allows at least in part a decoupling of their evolutionary pathways.

The decoupling of evolutionary pathways of *Hox* genes within different structures and tissues presents further evidence for the evolutionary malleability of *Hox* gene function and for the co-option of new functions for *Hox* genes (Duboule & Wilkins 1998).

Pattern and process in the conservation of *Hox* genes

Rostrocaudal gradient in conservation

When analysing the conservation of *Hox* genes, the most striking phenomenon is the strong rostrocaudal gradient in conservation among *Hox* genes along the anteroposterior axis (e.g. van der Hoeven et al 1996). This gradient can be found in the similarity of the DNA composition, as well as in the expression patterns between paralogous genes of the same group, the regularity in expression patterns of neighbouring genes along the same cluster, the differences between expression patterns of the same *Hox* genes in CNS and mesoderm, and the differences in expression patterns between species of different vertebrate classes. It is likely that there is a relationship between the lack of overlap in expression patterns and the degree of functional differentiation, after all, expression in a different part of the body will usually have a different effect on the phenotype. The rostrocaudal gradient of conservation is reflected in the degree of developmental control of structures of the vertebral axis, which is also significantly higher rostrally than caudally (Sofaer 1978, Schulz 1961).

To give an example of a rostrocaudal gradient in conservation, in the paraxial mesoderm of mice the gene expression areas of the rostral genes *Hoxb1*, *Hoxb2* and *Hoxa2* have the same rostral border (Prince et al 1998b), whereas the rostral borders of expression of *Hoxc13* and *Hoxd13* differ by seven somites and *Hoxb9* and *Hoxc9* by even 11 somites (cf. Peterson et al 1994, Kessel & Gruss 1991).

It is paradoxical that despite the strong rostral conservation, the most rostral part of the body, the head, is impressively variable within vertebrates, both with respect to the CNS and to the musculoskeletal parts. However, the strong rostral conservation in the genes is reflected in the almost complete lack of variation in the number of rhombomeres in vertebrates and in the number of cervical vertebrae in mammals.

Selective factors preserving conservation of Hox *gene expression: the case of the constant number of cervical vertebrae in mammals*

Mammals have a remarkably constant number of seven cervical vertebrae. The constancy of this number is not due to a lack of variation in vertebral number, because mutations for six, but also eight, cervical vertebrae do occur in mammalian species (e.g. Schulz 1961). Furthermore, transgenic mice with aberrant expression patterns of *Hox* gene groups 4 to 8 often have, surprisingly, six (and sometimes eight) cervical vertebrae (refs in Horan et al 1995, Galis 1999). There is, thus, strong stabilizing selection against these mutations in mammals, but not in birds and reptiles because cervical vertebral numbers vary greatly among species of these classes. The selective forces that act against variation in the number of cervical vertebrae in mammals are most likely: (i) an increased susceptibility for cancer and still births; and (ii) neuronal problems in the cervicothoracic area (Galis 1999).

Hox genes play an important role in this evolutionary constraint. In vertebrates *Hox* genes (amongst others) are involved in the specification of vertebrae identity (cervical versus thoracic in this case), in the development of the nervous system, and in mammals in the development of cancers that are associated with variations in cervical vertebral number. *Hoxb8* is of interest in this case. Rostral overexpression of *Hoxb8* in mice leads to six cervical vertebrae and to neuronal abnormalities (Charité et al 1994), whereas overexpression of *Hoxb8* in bone marrow leads to leukaemia (Kongsuwan et al 1989, Perkins & Cory 1993) and in fibroblasts to fibrosarcoma (Aberdam et al 1991). Of special interest are mice with mutations in Polycomb and Trithorax group genes. These mutants have shifts in *Hox* gene expression patterns, homeotic transformations along the entire vertebral axis (including a deviant number of cervical vertebrae of six or eight) and leukaemia (e.g. van der Lugt et al 1996, Akasaka et al 1996). Furthermore, embryonal cancers in humans are surprisingly often associated with cervical ribs (a partial or whole transformation of the seventh cervical vertebra to a thoracic vertebra; Schumacher et al 1992). A comparison of the medical, developmental and veterinary literature points to a coupling between functions of *Hox* genes that appears to be lacking in birds, reptiles and amphibians (Galis 1999), or at least has no apparent consequences when cervical vertebral number is changed.

A hypothesis which explains the emergence of a coupling of proliferation and axial patterning is that in mammals *Hox* genes have acquired a new function in proliferation; a function that is linked to the activity of *Hox* genes in the somitic mesoderm during the time when the positional information of the vertebrae is specified. This hypothesis is in agreement with the current notion that the development of complexity was accompanied and made possible by an increase in biological functions per regulatory gene (Duboule & Wilkins 1998). It is surprising that no decoupling of transcriptional control has evolved between these functions, since the increased susceptibility for, amongst others, cancer is clearly selectively disadvantageous. During neurulation there are many mutual interactions of neural crest, neural tube, notochord and somites (Gossler & Hřabe de Angelis 1998). These intense interactions during neurulation may be responsible for the conserved coupling of functions of *Hox* genes in mammals with, as a result, an increased risk of cancer, still births and neuronal changes when structural changes occur in the cervical region (Galis 1999). The constraint on the mammalian cervical vertebrae is, thus, due to internal selection.

This case shows that the key to the remarkable conservation of *Hox* genes in vertebrates may lie in a further understanding of the intricate and interactive processes that occur during neurulation and somitogenesis. It also shows that fundamental differences in *Hox* gene function have apparently evolved between species of different vertebrate classes.

Homology of *Hox* genes and vertebral regions

The vertebral column

A particularly promising model system for the establishment of homology between *Hox* genes and structures is the vertebral column. After all, conservation of expression patterns and functions of *Hox* genes is strongest along the anteroposterior axis, and the above-mentioned conservation in the cervical vertebrae adds to the expectation of meaningful interpretations. Gaunt (1994) and Burke et al (1995) have drawn attention to similarities in shifts of *Hox* gene expression and shifts in boundaries between vertebral regions in chickens and mice. Burke et al (1995) have claimed that in this case there is homology of *Hox* genes and vertebral regions and/or of position of the limbs for birds and mammals, and even *Xenopus* and zebrafish.

There are, however, reasons to expect important differences in the genetic regulation of vertebral regions between birds, mammals and other vertebrates. Changes of the vertebral column have always been important for the evolution of vertebrates. For instance, in the early history of tetrapods force transmission during locomotion changed drastically and had to be transmitted from the limbs to the vertebral column (Radinsky 1987). In early tetrapods the connection between the vertebral column and the small pelvic girdle was formed by a rib, and the force of the hindlimb was transmitted via this rib. Other examples of evolutionary importance are the countless changes in number of vertebrae that accompanied and accommodated shape changes. The important changes in functional demands of the axial skeleton lead us to expect extensive changes in the functioning of genes involved in the specification of identity (and number) of vertebrae. *Hox* genes are without a doubt involved in the specification of identity of vertebrae[2] (e.g. Kessel 1992) and thus in the determination of the boundaries between vertebral regions. Thus, it is important to investigate in more detail the claim about homology of structures and *Hox* genes in this case.

Is it possible to predict the vertebral boundaries on the basis of *Hox* gene expression patterns and current knowledge of *Hox* gene function? Gaunt (1994) and Burke et al (1995) have drawn attention to the position of the *Hoxc6* gene in the paraxial mesoderm, which has its most rostral border at approximately the future cervicothoracic boundary in chickens, geese and mice, despite the much larger number of cervical vertebrae in chickens. This is, indeed, a striking phenomenon, although it should be kept in mind for the generality of homology outside birds and mammals that in *Xenopus Hoxc6* is apparently not involved in axial patterning (Wright et al 1989). *Hoxc6* is likely to be involved in the specification of identity of vertebrae in that region in mice, since most *Hox* genes of groups 4 to 8 are involved in the specification of that boundary (Horan et al 1995). Shifts of the cervicothoracic boundary in a posterior direction (cervical ribs) occur particularly often in transgenic mice with loss- or gain-of-function mutations (*Hoxa4*, *Hoxd4*, *Hoxa5* and *Hoxa6* loss-of-function, and *Hoxb7* and *Hoxb8* gain-of-function, refs in Horan et al 1995, Galis 1999). This boundary is particularly susceptible to changes in *Hox* gene expression (Horan et al 1995). The expression patterns of most of these genes do not mark the boundary, and they do not all differ by six somites between mice and chickens, e.g. *Hoxb8* differs by only one somite (cf. Charité et al 1995, Lu et al 1997). To add to the complexity, *Hoxa5* influences the thoraco-lumbar boundary as well, because null mutants not only have a cervical rib (an anterior transformation) but also a lumbar rib (a posterior transformation). The same posterior shift of the thoraco-lumbar boundary is produced by caudal overexpression of *Hoxc6*; thus, the caudal border of gene expression is also of importance. Lumbar ribs are also found in *Hoxc8* loss-of-function mutants and in gain-of-function mutants of the same gene (references in Charité et al 1995). This puzzling phenomenon, that overexpression leads to the same phenotype as a null mutation, may be related to the existence of more than one

---

[2]It is puzzling that there is an almost complete lack of duplicated vertebrae in transgenic mice with mutated *Hox* genes, whereas duplications are common evolutionary changes. However, they are easily induced with retinoic acid (Kessel & Gruss 1991, Kessel 1992) which also modulates the expression of many *Hox* genes.

protein transcribed from a *Hox* gene. In *Xenopus, Hoxc6* is known to produce a long and a short protein with different expression areas and different functions (Wright et al 1989). Overexpression of the short protein produces the same phenotypic effect as loss-of-function of the long protein, and it differs from the effect of overexpression of the long protein. These proteins probably compete for the same binding sites, because they show antagonistic effects (the short protein has an inhibitory effect on the development of neurons expressing long *Hoxc6* proteins). *Hoxc6* in mice also produces two proteins (Shimeld et al 1993) and the expression patterns differ by three somites in the prevertebral column (thus, one protein of *Hoxc6* does not mark the cervicothoracic border). *Hoxb6* and *Hoxb3* in mice also are known to have several transcripts (3 and 6, refs in Shimeld et al 1993).

In addition to the complexity of interactions described above, there are quantitative effects (the phenotypic effect depends on the number of active paralogous or even non-paralogous genes), differences in the timing of *Hox* gene expression and extensive auto- and cross-regulation of *Hox* genes involved in the development of the vertebral column. Based on current knowledge it is, therefore, far from possible to predict the boundary between different vertebral regions on the basis of gene expression patterns. Without denying the significance of *Hox* gene expression shifts, I want to conclude that the simple assignment of homology between expression patterns of *Hox* genes and boundaries of vertebral regions is not feasible at the moment. Such an assignment of homology does not do justice to the current knowledge of complex interactions involved in *Hox* gene regulation. Certainly, more is involved than parallel shifts of gene expression patterns. This makes intuitive sense because shifts of vertebral boundaries have not occurred in isolation, but have been accompanied by extensive remodelling of the vertebral column. The functional differentiation of *Hox* genes necessary for this remodelling will have required readjustment of interacting paralogous and non-paralogous genes.

The second hypothesis of Burke et al (1995) on the causal relationship between the expression of *Hoxc6* (and other *Hox* genes) in the paraxial mesoderm and limb position, has been superseded by recent studies. *Hox* gene function in the lateral plate mesoderm is involved with the specification of limb position independently of *Hox* gene expression in the paraxial mesoderm (Cohn et al 1997, see also Zakany et al 1997). Furthermore, *Hox* gene expression in the paraxial mesoderm differs considerably from that in the lateral plate mesoderm (cf. *Hox9* genes in Kessel & Gruss 1991, Cohn et al 1997).

Ultimately, as an evolutionary biologist, one wants to understand both developmental and evolutionary transformations. The assignment of homology is a useful tool in the analysis of transformation, but the complexity of interactions involved should lead one to accept that often there is no simple conclusion with respect to the homology of structures, nor to the relationship between genetic identity and structural identity (Müller & Wagner 1996, Bolker & Raff 1996, Galis 1996). The more structures are used for different functions, the more selection on developmental processes and on genes moulding the character obscure the history of the evolutionary process and, more importantly, change the developmental process to such an extent that the establishment of homology becomes meaningless.

In the case of the homology between *Hox* genes within vertebrates it is clear that, after the initial enthusiasm about the conservation and emphasis on similarities, it is now time for a thorough investigation of the differences between organisms. This is necessary to understand the development of the fundamental differences that have arisen in vertebrate species. The similarities in gene expression patterns can be helpful in indicating where one should look for differences, e.g. the focus on the interactions of *Hoxc6* with other relevant genes in the case of the shift of the cervicothoracic boundary. A process-oriented approach will be more productive than a pattern-oriented approach. The search for homologies will continue to provide important insights as long as the relationship with the evolutionary process is not lost.

Acknowledgements

I thank Adam Wilkins and Hans Metz for many clarifying and stimulating discussions and helpful comments.

References

Aberdam D, Negreanu V, Sachs L, Blatt C 1991 The oncogenic potential of an activated *Hox-2.4* homeobox gene in mouse fibroblasts. Mol Cell Biol 11:554–557

Akasaka T, Kanno M, Balling R, Mieza MA, Taniguchi M, Koseki H 1996 A role for *mel-18*, a Polycomb group-related vertebrate gene, during the anteroposterior specification of the axial skeleton. Development 122:1513–1522

Aparicio S, Hawker K, Cottage A et al 1997 Organization of the *Fugu rubripes hox* clusters: evidence for continuing evolution of vertebrate *Hox* complexes. Nature Genet 16:79–83

Bolker J, Raff RA 1996 Developmental genetics and traditional homology. BioEssays 18:489–494

Burke AC, Nelson CE, Morgan BA, Tabin C 1995 *Hox* genes and the evolution of vertebrate axial morphology. Development 121:333–346

Charité J, Graaff WD, Shen S, Deschamps J 1994 Ectopic expression of *Hoxb-8* causes duplication of the ZPA in the forelimb and homeotic transformation of axial structures. Cell 78:589–601

Cohn MJ, Patel K, Krumlauf R, Wilkinson DG, Clarke JDW, Tickle C 1997 *Hox9* genes and vertebrate limb specification. Nature 387:97–101

Duboule D, Wilkins AS 1998 The evolution of 'bricolage'. Trends Genet 14:54–59

Galis F 1996 The evolution of insects and vertebrates: homeobox genes and homology. Trends Ecol Evol 11:402–403

Galis F 1999 Why do almost all mammals have 7 cervical vertebrae? Developmental constraints, *Hox* genes and cancer. Mol Dev Evol, in press

Galis F, Metz JAJ 1998 Why are there so many cichlid species? Trends Evol Ecol 13:1–2

Gaunt SJ 1994 Conservation in the *Hox* code during morphological evolution. Int J Dev Biol 38:549–552

Godsave S, Dekker E-J, Holling T, Pannese M, Boncinelli E, Durston A1994 Expression patterns of *Hoxb* genes in the *Xenopus* embryo suggest roles in anteroposterior specification of the hindbrain and in dorsoventral pattering of the mesoderm. Dev Biol 166:465–476

Godwin AR, Capeechi MR 1998 *Hoxc13* mutant mice lack external hair. Genes Devel 12:11–20

Gossler, Hřabe de Angelis 1998 Somitogenesis. Curr Top Dev Biol 38:225–287

Horan GSB, Nagy Kovacs E, Behringer RR, Featherstone MS 1995 Mutations in paralogous *Hox* genes result in overlapping homeotic transformations of the axial skeleton: evidence for unique and redundant function. Dev Biol 169:359–372

Janvier P 1996 Early vertebrates. Clarendon Press, Oxford

Kessel M 1992 Respecification of vertebral identities by retinoic acid. Development 115:487–501

Kessel M, Gruss P 1991 Homeotic transformations of murine vertebrae and concomitant alteration of *Hox* codes induced by retinoic acid. Cell 67:89–104

Knittel T, Kessel M, Kim HM, Gruss P 1995 A conserved enhancer of the human and murine *Hoxa-7* gene specifies the anterior boundary of expression during embryonal development. Development 121:1077–1088

Kongsuwan K, Allen J, Adams JM 1989 Expression of *Hox-2.4* homeobox gene directed by proviral insertion in a myeloid leukemia. Nucleic Acids Res 17:1881–1892

Krumlauf R 1992 Evolution of the vertebrate *Hox* homeobox genes. BioEssays 14:245–252

Lu H-C, Revelli J-P, Goering L, Thaller C, Eichele G 1997 Retinoid signaling is required for the establishment of a ZPA and for the expression of *Hoxb-8*, a mediator of ZPA formation. Development 124:1643–1651

Lufkin T 1996 Transcriptional control of *Hox* genes in the vertebrate nervous system. Curr Opin Genet Devel 6: 575–580

Müller GB, Wagner GP 1996 Homology, *Hox* genes, and developmental integration. Am Zool 36:4–13

Perkins AC, Cory S 1993 Conditional immortalization of mouse myelomonocytic, megakaryocytic and mast cell progenitors by the *Hox-2.4* homeobox gene. EMBO J 12:3835–3846

Peterson RL, Papenbrock T, Davda MM, Awgulewitsch A 1994 The murine *Hoxc* cluster contains five neighbouring AbdB-related *Hox* genes that show unique spatially coordinated expression in posterior embryonic regions. Mech Dev 47: 253–260

Prince VE, Joly L, Ekker M, Ho RK 1998a Zebrafish *hox* genes: genomic organization and modified colinear expression patterns in the trunk. Development 125: 407–420

Prince VE, Moens CB, Kimmel CB, Ho RK 1998b Zebrafish *hox* genes: expression in the hindbrain region of wild-type and mutants of the segmentation gene, *valentino*. Development 125:393–406

Radinsky LB 1987 The evolution of vertebrate design. University of Chicago Press, Chicago, IL

Rijli FM, Chambon P 1997 Genetic interactions of *Hox* genes in limb development: learning from compound mutants. Curr Opinion Genet Devel 7:481–487

Rogina B, Coelho C, Kosher RA, Upholt WB 1992 The pattern of expression of the chicken homolog of *HOX*1I in the developing limb suggests a possible role in the ectodermal inhibition of chondrogenesis. Dev Dyn 193:92–101

Schulz AH 1961 Vertebral column and thorax. In: Hofer H, Schulz AH, Starck D (eds) Primatologia, handbook of primatology. S Karger AG, Basel, p 5/1–5/66

Schumacher R, Mai A, Gutjahr P 1992 Association of rib anomalies and malignancy in childhood. Eur J Pediatr 151:432–434

Shimeld SM, Gaunt SJ, Coletta PL, Geada AMC, Sharpe PT 1993 Spatial localisation of transcripts of the *Hox-c6* gene. J Anat 183:515–523

Sofaer JA 1978 Morphogenetic influences and patterns of developmental stability in the mouse vertebral column. In: Butler PM, Joysey KA (eds) Development, function and evolution of teeth. Academic Press, London, p 215–227

van der Hoeven F, Sordino P, Fraudeau N, lzpisfla-Belmonte JC, Duboule D 1996 Teleost *HoxD and HoxA* genes: comparison with tetrapods and functional evolution of the *HOXD* complex. Mech Dev 54:9–21

van der Lugt NMT, Alkema M, Berns A, Deschamps J 1996 The polycomb-group homolog *Bmi-1* is a regulator of murine *Hox* gene expression. Mech Dev 58:153–164

Wright CVE, Cho KWY, Hardwicke J, Collins RH, DeRobertis EM 1989 Interference with function of a homeobox gene in *Xenopus* embryos produces malformations of the anterior spinal cord. Cell 59:81–93

Xue Z, Xue XJ, Le Douarin NM 1993 *Quox*-1, an *Antp*-like homeobox gene of the avian embryo: a developmental study using a *Quox*-1-specific antiserum. Mech Dev 43:149–158

Xue Z, Ziller C, Xue XJ 1998 Quox 1 homeobox protein is expressed in postmitotic sensory neurons of dorsal root ganglia. Devel Brain Res 105:59–66

Zakany J, Gérard M, Favier B, Duboule D 1997 Deletion of a *HoxD* enhancer induces transcriptional heterochrony leading to transposition of the sacrum. EMBO J 16:4393–4402

Zimmermann F, Rich IN 1997 Mammalian homeobox B6 expression can be correlated with erythropoietin production sites and erythropoiesis during development, but not with hematopoietic or nonhematopoietic stem-cell populations. Blood 89:2723–2735

DISCUSSION

Akam: A word of caution about the assumption that there is a linkage between the misregulation of the *Hox* genes and cancer. I would guess that many of the naturally occurring variants in *Hox* gene expression are not due to mutations in the *Hox* genes themselves, but are due to mutations in the Polycomb or Trithorax group genes, which will deregulate not only the *Hox* genes, but also many other targets. For example, one of the Trithorax group genes is the *Drosophila* homologue of vertebrate GAGA factor, which is a widely used transcription factor. Therefore, the fact that you see a pairing of these phenotypes in individuals with embryonic cancer could simply be due to the deregulation of many genes, some of which lead to the cancer phenotype. The deregulation of the *Hox* genes leads to the skeletal phenotypes, but the *Hox* gene mis-expression may be independent of the cancer phenotype.

Galis: I didn't want to imply that it must be a mutation in the *Hox* genes, but rather that it seems likely. Many geneticists who work on cancer are interested in the abnormalities of *Hox* gene expression because they think it plays an important part in the normal and abnormal proliferation of cells. At the same time, mutations of *Hox* genes are involved in vertebral abnormalities, so it is likely that they provide the link between embryonal cancers and the congenital abnormality of a cervical rib. In addition, these embryonal cancers are not usually familiar, and in that case they are not interesting to discuss because they are not able to pass this character to the next generation. However, a small percentage are familial. The occurrence of congenital abnormalities (such as cervical ribs) is much higher in those cases.

Akam: Unless you have a high cancer incidence in *Hox* gene mutants themselves, it is unwise to assume that this link is causal.

Galis: Overexpression of *Hoxb8* has been known to cause cancer and induce cervical ribs.

Wagner: The important point is the genetic correlation between the morphological phenotype and the pathological phenotype.

Akam: In flies by far the most common cause of ectopic *Hox* gene expression is a mutation in the Polycomb or Trithorax group genes.

Wagner: But as long as the cancer and the axial differentiation share a developmental pathway there will be a genetic correlation.

Tautz: I would like to bring up a more general point that you raised, namely how much variation is there in *Hox* gene expression patterns in different taxa? On the one hand, it is always emphasized how much similarity there is between *Hox* expression patterns of such distantly related species as *Drosophila* and mouse. On the other hand, one can find notable differences in expression patterns already among crustaceans (Averof & Patel 1997). There could be a lot of evolutionary flexibility in these patterns.

Galis: There is a lot of conservation, but there is also a lot of variation.

Tautz: But why should this be so?

Wray: There is a natural tendency to seize upon surprising similarities and to filter out the differences. If we look at two taxa, we often find something that's similar, and then as we study more taxa we find that this is not one of the fundamental, similar elements.

Akam: I would like to ask a more general question to the palaeontologists. You said you were unhappy about using gene expression patterns to define homologies between different vertebrae.

Galis: Yes, at this moment in time for these specific examples.

Akam: Quite rightly so. But what criteria would you use to define homology between different regions of the vertebral column, if you don't accept that the thorax of one animal is homologous to the thorax of another because those vertebrae carry ribs?

Galis: In the past some people have used a difference in the articulation of the 10th lumbar vertebra as an indication of the lumbosacral transition. Thus, although in general there is consensus about the lumbosacral transition, one could argue about the position of the transition for morphological reasons.

Akam: Would I make enemies if I said the first thoracic vertebra is homologous in all tetrapods?

Galis: Sometimes vertebrae are not fused to the sternum, so then it becomes arbitrary whether you define it as a first thoracic vertebra or as a cervical vertebra.

Rudolf Raff: Surely, zebrafish and *Xenopus* do not have definitive cervical or thoracic lumbar vertebrae.

Galis: No, this claim was only for birds and mammals. The alternative hypothesis was with respect to the position of the limbs, which applies to *Xenopus* and zebrafish. The interesting observation in this case is that the *Hoxc6* gene does not seem to be involved in the patterning of the skeletal axis in *Xenopus*. This hypothesis can be discarded because the *Hox* gene function in the vertebral column is separate from the limbs, and the *Hox* gene expression in the lateral plate mesoderm is more important. This is not the same as in the preaxial mesoderm. This illustrates the danger of simplifying the relationship between *Hox* gene expression patterns and structure.

Wagner: Only if a thorax is present. The thorax in that sense may be a derived character, and once it is established then you might have a strong association between *Hoxc6* expression, for example, and limb formation. But in zebrafish and *Xenopus* this doesn't have to be true.

Hall: This brings up the question of how closely related do organisms have to be for us to be able to make statements about homology at particular levels. In terms of the cervical vertebrae, we're talking about some closely related animals about which we can make comparisons, and outside that we can't. How does one make those decisions, other than on an arbitrary basis?

Panchen: One makes arbitrary decisions that are more likely to be real the more closely related the animals are that one is comparing.

Müller: I have a comment and a question. I believe your examples of how changes in completely different genes lead to the same morphological result, i.e. shifts in morphology of vertebrae with or without ribs, nicely illustrate the propensity of developmental systems to generate similar structures under different influences. My question is, why do you find it necessary to invoke selection for maintaining the homology of a region like the cervical vertebrae? Couldn't it just be that in mammals morphological integration in that area has reached a point in which it is impossible for many reasons, developmental, structural and functional, to change the number of vertebrae?

Galis: Mutations changing the number of cervical vertebrae do occur. It's not that there is a lack of variation in cervical rib number. Cervical ribs also occur infrequently in other mammals, so the variation is there, but the character of a different number of cervical vertebrae never becomes established in populations.

Peter Holland: There is a difference between the natural situation in birds and the experimental situation in mice. The former is a change in segment number, the latter is a homeotic transformation. In the variant human conditions, what's actually happening is also a homeotic transformation.

Galis: An increase in the number of vertebrae occurs as a result of a duplication event, but at the moment we don't know anything about the genetic basis of this.

Homeotic transformations are also important in evolution as duplications. Changes in the number of vertebrae of a certain vertebral region can be the result of homeotic changes as well as of duplications.

Peter Holland: I still think the latter are a change in number of vertebrae, not homeosis. We don't know the genetic basis of change in segment number, so we can't say whether there is natural variation at the level of *Hox* gene expression.

Galis: But homeotic changes also occur in humans that lead to an increased number of vertebrae in a certain region, for example in the sacral vertebrae. Duplication may lead to the same effect. Whenever we see an aberration in the number of cervical vertebrae, it is always combined with a high rate of lethality.

Peter Holland: But those specific cases represent unusual types of changes in cervical vertebrae morphology, i.e. they are homeotic transformations. They are not the same as the differences between a giraffe and a swan, which are not due to homeotic transformations, but rather to a vast increase in the number of vertebrae in a particular region of the vertebral column.

Galis: In the case of swans there must have been duplications, but many birds have around 13, 14 or 15 cervical vertebrae, and the number of thoracic vertebrae are particularly small.

Rudolf Raff: Do we know that the duplications are caused by the *Hox* genes, or is another genetic mechanism involved?

Wray: It is not known. It is possible that retinoic acid is involved.

Galis: I have looked and I have asked people whether they know of any duplication mutants in mice, but no one has found any. Then, I finally found duplications in experiments in which retinoic acid was applied early in ontogeny (Kessel 1992, Kessel & Gruss 1991). There is another gene that seems to inhibit retinoic acid, *ActRIIB*, and mice with null mutations of this gene also had duplications of vertebrae and altered expression of *Hox* genes (Oh & Li 1997).

Rudolf Raff: Retinoic acid may be affecting another set of genes in addition to the *Hox* genes, or another set of genes may be involved in the actual evolutionary duplications and this is not the set that is affected by retinoic acid.

References

Averof M, Patel N 1997 Crustacean appendage evolution associated with changes in *Hox* gene expression. Nature 388:682–686

Kessel M 1992 Respecification of vertebral identities by retinoic acid. Development 115:487–501

Kessel M, Gruss P 1991 Homeotic transformations of murine vertebrae and concomitant alteration of *Hox* codes induced by retinoic acid. Cell 67:89–104

Oh SP, Li E 1997 The signaling pathway mediated by the type IIB activin receptor controls axial patterning and lateral symmetry in the mouse. Genes Dev 11:1812–1826

Developmental basis of limb homology in urodeles: heterochronic evidence from the primitive hynobiid family

J. R. Hinchliffe and E. I. Vorobyeva*

*Institute of Biological Sciences, University of Wales, Aberystwyth, Wales, SY23 3DA, UK, and *Institute of Ecology and Evolution, 33 Leninsky Prospect, Moscow 117071, Russia*

Abstract. The vertebrate limb is a classic example of homology, long assumed to be underpinned by a developmental 'bauplan' of the type proposed in the Shubin/Alberch branching and segmenting model. In the anuran/amniote pattern skeletogenesis proceeds in a proximodistal direction with digits forming from the posterior to the anterior. But in free-living larvae of 'advanced' urodeles, the pattern of skeletogenesis is distinctly different with digits 1 and 2 and the basal commune developing early, in an anterior/distal position. This different pattern is cited as evidence for a diphyletic theory of tetrapod evolution. Reassessing this problem, we analysed the pattern of early skeletogenesis of three genera (*Salamandrella*, *Ranodon*, *Onychodactylus*) of the 'basal' family of hynobiids, using immunofluorescence to localize chondroitin-6-SO_4 in *Salamandrella*. Here the developmental sequence was more proximodistal (intermedium preceding basal commune; early formation of the digital arch). This pattern, also found in direct developing urodeles such as *Bolitoglossa subpalmata*, resembled that in anurans/amniotes. Uniquely amongst tetrapods, urodeles use their developing limbs for locomotion. We attribute the unusual pattern in 'advanced' urodeles to adaptive modification of the developing limb. Differences in the pattern between 'basic' and 'advanced' urodeles and between urodeles and anuran/amniotes are interpreted as heterochronic within an overall single tetrapod developmental bauplan.

1999 Homology. Wiley, Chichester (Novartis Foundation Symposium 222) p 95–109

Richard Owen (1849) in illustrating his newly defined concept of homology considered the vertebrate limb as a classic example of the fundamental similarity which this concept entailed. Later, for Darwin in '*The origin of species*', the limb pentadactyl pattern illustrated the principle of adaptive radiation and of descent from a common ancestor (Darwin 1859). Today, much additional fossil (and developmental) data are available, and in early tetrapods and their lobe-fin fish ancestors the homology of the limb skeletal parts is clear in the proximal parts, including stylopod (humerus/femur), zeugopod (radius/ulna, tibia/fibula) and the intermedium (Vorobyeva & Hinchliffe 1996a). The more distal parts (mesopod and digits) have been the subject of discussion from the early part of the century (Sewertzov 1908, Schmalhausen 1915, Holmgren 1933) to recent times (Jarvik 1980, Thomson 1968, Hinchliffe & Griffiths 1983, Hinchliffe 1994, Shubin 1991, 1995, Duboule 1994, Coates 1995). However, the consensus today is that the distal elements, and especially the digits of the tetrapod limb, are not homologous with the radials of the paired lobe-fin of Devonian fishes such as *Eusthenopteron* and *Panderichthys*, and that carpal/tarsal elements and the digits are probable neomorphs.

This viewpoint is connected with the notion that homology in morphology is connected with structural and developmental stabilization. At the same time, neomorphic structures demonstrate much variability between different taxa, and they provide a good subject for analysis of the part played by heterochrony and adaptive modification in the process of diversification. Such neomorphic structures are frequently most unstable in the groups closest to the ancestral condition. Thus, it is not surprising that in Palaeozoic times the first aquatic tetrapods were polydactylous (Coates 1995, Gould 1994), suggesting that the underlying morphogenetic processes were highly plastic. This concept is supported by evidence in extant forms from experimental embryology that it is more difficult to disturb proximal than distal limb skeletal development—the 'proximal stability, distal instability' rule (Hinchliffe 1994). Such plasticity appears to be retained among recent amphibians at both inter- and intraspecific levels (Blanco & Alberch 1992, Shubin & Wake 1996).

Molecular evidence has now been brought to bear on this problem. In particular, *Hox* gene expression patterns during development have been analysed, and they appear to underpin the concept of proximal and distal differences in stability (Duboule 1994, Coates 1995). Although the proximal parts of fish paired fin buds and tetrapod limb buds share common expression patterns, the more recently evolved distal limb parts appear governed by specific expression patterns in the limb bud not found distally in fish fin buds.

Is urodele developmental patterning of the limb skeleton unique among tetrapods?

In this analysis we describe new evidence from urodele (tailed amphibia) embryology that relates to limb homology and its developmental base. Regarding the limb homology of the different tetrapod groups, the main questions to be asked are:

(1) How correct is the interpretation of Holmgren (1933) and Shubin & Alberch (1986) which separates the urodele developmental pattern from the common developmental bauplan of non-urodelan (anuran/amniote) tetrapods?

(2) Which ancestral features are retained in urodeles?
(3) Which are the urodele groups closest to the ancestral condition?
(4) What features of developing limbs represent a common urodele pattern and what parts are subject to variation in different taxa in particular through heterochronic differences?
(5) What is the relationship between the developmental variability of urodele limbs and their biodiversity as represented by their different life cycles and ecologies and their adaptations?

Special attention is currently being paid to variation in limb patterning in urodeles (Shubin & Wake 1996, Blanco & Alberch 1992, Alberch & Blanco 1996) and especially to patterning in salamanders with direct development such as those plethodontids which have secondarily lost their metamorphosis (Shubin 1995, Wake & Hanken 1996). Among these is *Bolitoglossa palmata*, which has some developmental similarities with amniotes (Wake & Hanken 1996), in contrast with larval advanced salamanders, such as *Triturus marmoratus* (Blanco & Alberch 1992). In the context of homology, these similarities may represent the retention of some ancestral characters common to all tetrapods and whose underlying morphogenesis has been retained in latent form.

Limb development in the primitive hynobiid family of urodeles

The Asian salamander family Hynobiidae is considered to possess the most primitive characters among recent caudata or tailed amphibia (Duellman & Trueb 1986, Kuzmin 1995). This is on the basis of their external fertilization and specific morphological and karyological features. Members of this family are thus of particular interest in relation to questions of limb homology and evolution, since probably they represent the most primitive extant tetrapods.

We compared in some detail development of the larval limbs (especially skeletal development) in two hynobiid species (*Salamandrella keyserlingii* and *Ranodon sibiricus*) and rather more superficially in a third, *Onychodactylus fisheri*, for which only premetamorphic stages were available. *Salamandrella* represents a more primitive hynobiid group and has a wide distribution in Central Asia and Eastern European Russia. It belongs ecologically to the limnophilous type with pond-living larvae (Fig. 1). The other two species are more advanced hynobiids and their larvae are of the rheophilous or stream-living type. *Ranodon* has a narrow distribution in the mountain region of north Kazakhstan while *Onychodactylus* lives in mountain streams in the southern Far East.

The methods we used were a mix of classical histology and modern immunohistochemistry. Classical histological methods are sometimes not precise enough for mapping the early skeletogenic processes. Immunofluorescence

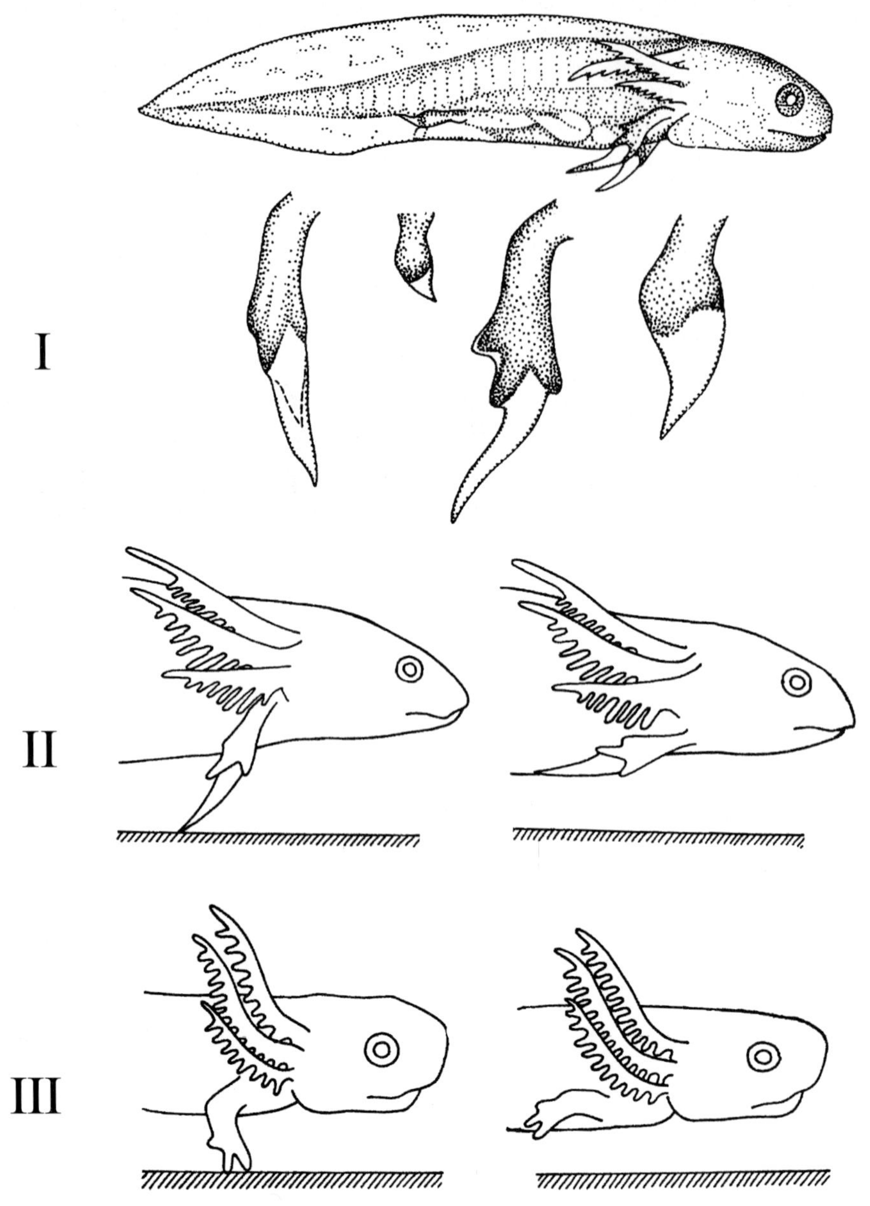

FIG. 1. Free-living larvae of urodeles use their developing limbs for locomotion. Developing limbs in a larva of *Salamandrella* (I). Fore- (large, 2–3 digit stage) and hindlimbs of this and an earlier stage are shown (length = 16–23 mm, from Sitina et al 1987). Separating digits 1–2 is a fin used initially to contact the substrate (II). Axolotls have larvae that use their developing forelimb digits 1 and 2 to contact the substrate (III). Both species use 'paddling' forward movement (II, III).

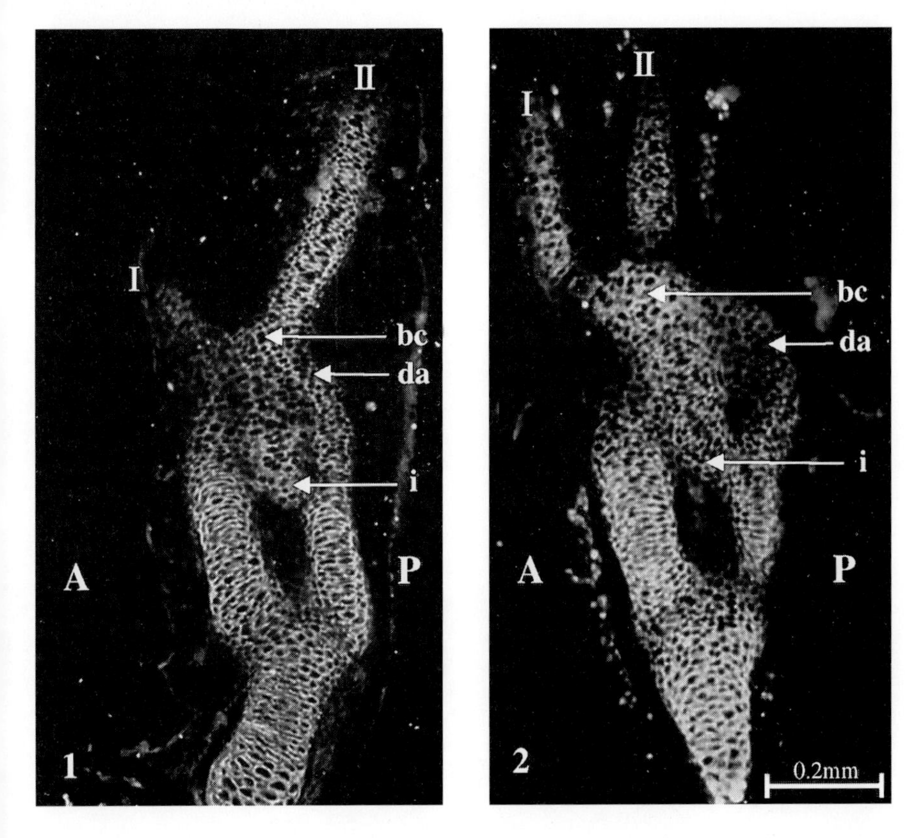

FIG. 2. Immunofluorescent sections of limbs of *Salamandrella* (1) and axolotl (2). Chondroitin-6-SO_4 distribution is shown. This technique maps the developing pattern of pre-chondrogenic and early limb cartilages, enabling differences in developmental timing to be clearly identified. A, anterior; bc, basal commune; da, digital arch; i, intermedium; P, posterior. I and II represent digit number.

staining in principle provides a more precise method and we found that an anti-chondroitin-6-SO_4 antibody (3B3 from B. Caterson, Cardiff, UK) proved useful in mapping pre-chondrogenic patterns in urodeles. We used this method for *Salamandrella* (with the axolotl *Ambystoma* for comparison, see Fig. 2), although not for the other hynobiid species as we did not have freshly fixed material on which to do immunochemical analysis.

Limb development in these hynobiid genera shows some common features (Vorobyeva et al 1997). They are characterized by early differentiation of the intermedium in comparison with the basal commune; early bifurcation by the ulna and fibula; early connection of the postaxial branch with the digital arch condensation; presence of two centralia; separate condensations for tarsal distal 4

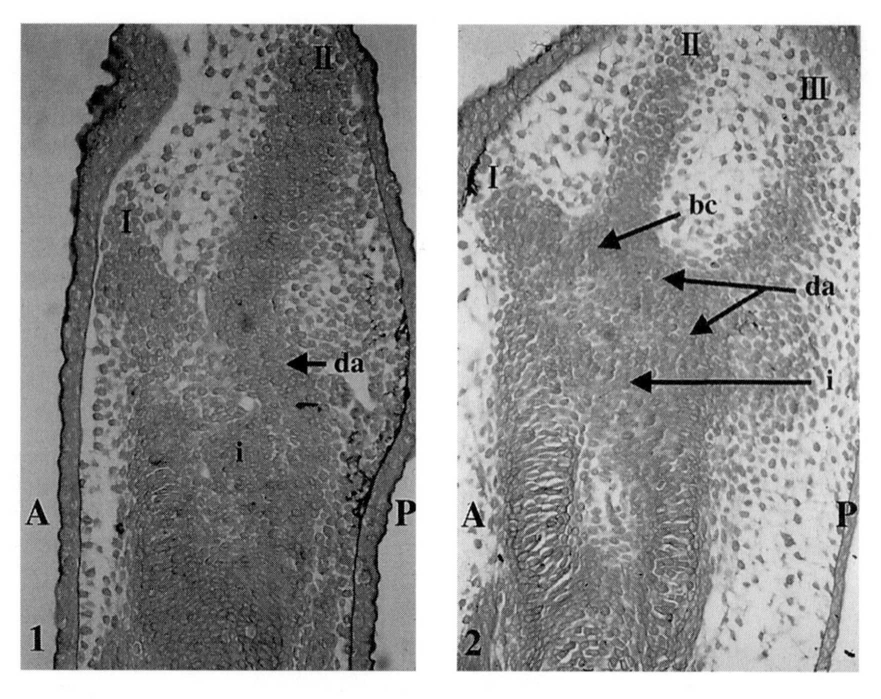

FIG. 3. Classical histology of *Ranodon* limb buds (1 & 2) illustrating proximodistal development including early appearance of intermedium (i) relative to the basal commune (bc) and early differentiation of the digital arch (da). A, anterior: P, posterior. I–III represent digit number.

and distal 5 (in *Ranodon*; *Salamandrella* has only four hind digits); and early differentiation and rapid segmentation of the preaxial branch (Figs 2, 3 & 4).

Hynobiid limb patterning compared with other urodeles

Some of the features of limb development in hynobiids can be considered as common to larval salamanders as a whole. Such features are: fork-like stylopod–zeugopod bifurcation; proximodistal differentiation of the preaxial arch (both probably sarcopterygian synapomorphies); organization of the mesopod into three columns; early differentiation of the pre-axial branch; early formation anteriorly of the first two digits and the associated basal commune (Fig. 4Biii); and anterior to posterior differentiation of the distalia.

At the same time these hynobiids display a number of features that make it possible provisionally to classify it as primitive (close to the ancestral condition). Among these are supernumerary condensations in the carpal/tarsal region. Some of

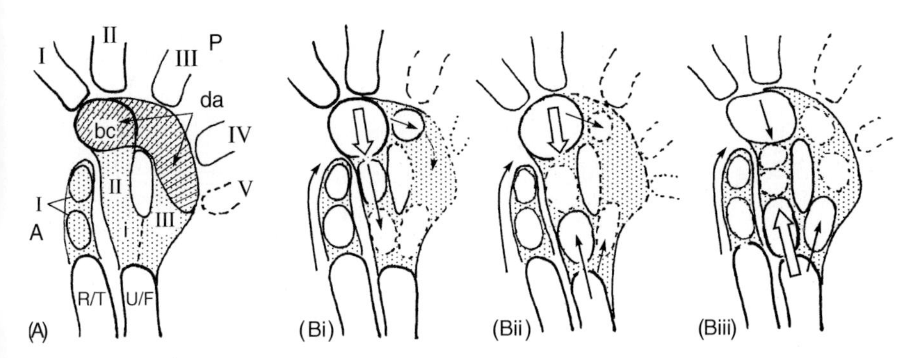

FIG. 4. (A) Schematic diagram of the common features of the pattern of the early developing limb skeleton in urodeles with non-reduced limbs. Note the three 'columns' (I, II and III, dotted), the 'digital arch' (da, hatched), basal commune (bc) and the anteroposterior sequence of digit development. A, anterior; F, fibula; i, intermedium; P, posterior; R, radius; T, tibia; U, ulna. I–V at top represent digit number. (B) Comparison of the pattern and sequence (graded arrows) of chondrogenesis in the tarsus of 'advanced' (i, *Triturus*; ii, axolotl) and basic urodeles (iii, *Salamandrella*).

these are present as interpopulation variations, e.g. presence of more than two centralia (mediale), prehallux, postminimus (Borkhvardt & Ivashintsova 1993). Others are standard features, such as the two centralia retained in the adult (except *Onychodactylus*) and the separate distalia 4 and 5 in the hindlimb (*Ranodon*, but not *Salamandrella* which has only four digits). All these features are characteristic of Palaeozoic temnospondyls such as *Trematops*. A feature of the evolution of the advanced urodeles such as *Triturus* is the reduction in the number of skeletal elements (one centralia only and fused distalia 4–5, see Shubin & Wake 1996).

A major difference between hynobiids and advanced salamanders is in the rate of development of the elements of the median column (Figs 2 & 4). In *Triturus* and *Ambystoma* the basal commune differentiates earlier than the intermedium and in *Triturus* the centrale also differentiates before the intermedium, and thus here the median column differentiates distoproximally. However, in *Ambystoma* the centrale is the last median column element to differentiate. Hynobiids have proximodistal differentiation of some median column elements, e.g. the early formation of the intermedium in relation to the basal commune (Figs 2 & 4Biii). There are also important differences in the timing of differentiation of the digital arch and the ulnare/fibulare. In *Triturus* this is relatively late and its elements develop proximally in an anteroposterior direction, whereas in the hynobiids it develops earlier and its elements form more or less at the same time (Figs 2, 3, 4 & 5).

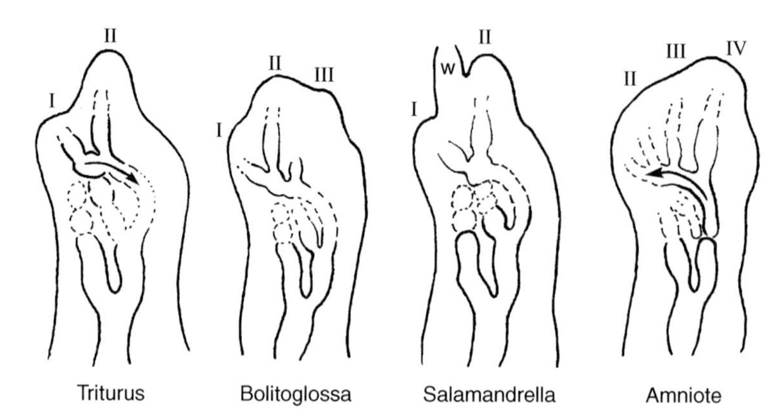

FIG. 5. Schematic diagram of developing urodele limbs, compared with the amniote pattern (based on Shubin & Wake 1996). Note that within an overall generally similar pattern of carpus/tarsus and metapod there is a spectrum from marked precocious early anterior and distal differentiation (*Triturus marmoratus*) to a more proximodistal differentiation and earlier appearance of the digital arch (basal and direct developing forms such as *Salamandrella* and *Bolitoglossa*) closer to the amniote pattern. Differences can be interpreted as heterochronic within an overall tetrapod developmental bauplan. w, web. I–IV represent digit number.

Comparison of the pattern of limb development in hynobiids with that in salamanders with direct development reveals some homoplasy. For example, *Bolitoglossa subpalmata* has some features such as early formation of the digital arch and postaxial column (Fig. 5, as in *Ranodon* and *Salamandrella*), described by Marks et al (1998) as 'amniote/anuran like'. *Triturus* in contrast to *Bolitoglossa* is characterized by late differentiation of the ulna/fibula bifurcation (Fig. 5). *Bolitoglossa* differs also from advanced larval salamanders with only a weak anteroposterior sequence of digit differentiation and (in common with hynobiids) early differentiation of the intermedium.

Developmental caenogenetic or heterochronic features that are adaptive

In this account we have focused mainly on changes in the timing of limb skeletogenic events (especially chondrogenesis) in different primitive and advanced urodele groups. It is difficult to identify many of these differences as specifically adaptive, e.g. early or late development of the digital arch or the sequence of condensation in the median column. However, during urodele limb development some features appear to be adaptive and relate to the different ecological niches of the species. Some are caenogenetic, such as the temporary web between digits 1 and 2 in *Salamandrella* (Fig. 1), a probable floating adaptation (Vorobyeva & Hinchliffe 1996b) in a pond-type species. Others are heterochronic, such as the early osteogenesis of digits and early claw formation in *Ranodon* and *Onychodactylus*. These have rheophilic or stream-living larvae that use their strengthened ossified digits to maintain a grip. Other related adaptations such as specialized tendon insertions to the digit tips are known. But rheophilic adaptations (Gregory 1996) are found regardless of the advanced or primitive status of the species and once again underline the importance of homoplasy in urodele evolution.

Finally, it is worth suggesting that in urodeles as a whole the generally precocious development of the digit 1–2 and basal commune complex is an adaptive heterochronic modification of the tetrapod bauplan, connected with limb use during larval development (Fig. 1). This modification is well marked in advanced salamanders but is less developed in hynobiids and in the direct developers. In the latter, selection for this is presumably relaxed, and the anuran/amniote bauplan prevails.

Conclusions

In general, we can see that the developing patterns of limb development in urodeles display a wide spectrum of diversification, in which primitive and advanced features present a mosaic picture, the result mainly of differences in the rate of condensation of different elements in the three columns (in themselves a stable feature in urodeles). However, the key role in these differences is played by the timing of appearance of the basal commune/digit 1 and 2 complex, which is a novel feature of urodeles in comparison with other tetrapods. The early appearance of this complex may have an adaptive character related to the fact that most larvae use their limb buds for locomotion during their development (Fig. 1).

The diversity of developmental processes supports the idea that:

(1) The carpus/tarsus region is a novelty in tetrapod evolution.
(2) The urodele condition is secondary in comparison with other tetrapods and is connected with the larval adaptation of the limb, a feature absent in other tetrapods.
(3) Urodeles display a spectrum of limb skeletal patterning, with hynobiids and direct-developing forms showing most similarities with those of anuran amniotes, while the advanced (free-living) forms show least.
(4) Rather than being seen as an exception, urodeles can now be fitted into a general tetrapod developmental bauplan for the limb (Hinchliffe 1991). This includes a stable branching stylopod and zeugopod. Distally, there is a preaxial segmenting ray and post-axially a 'digital arch' from which branch the metapods of the digits. However, the position of segmentation of the condensations (precursors of the definitive skeletal elements) differs in the

various taxa and the timing of events does not necessarily follow a strict proximodistal sequence. As demonstrated also by Alberch's studies on neoteny as the basis of both anuran and urodele limb variants, heterochrony is thus a critical factor in urodele limb evolution.

Acknowledgements

This research was supported by the Royal Society (Joint Project) and the Russian Foundation for Fundamental Research. We received valued assistance with immunofluorescence in Aberystwyth from Stephen Wade, Tracey Staughton and Patrick Tidball (Nuffield Foundation bursary); and with histology in Moscow from T. P. Antipenkova. The gift of the monoclonal antibody 3B3 from B. Caterson (University of Cardiff) is gratefully acknowledged.

References

Alberch P, Blanco MJ 1996 Evolutionary patterns in ontogenetic transformation: from laws to regularities. Int J Dev Biol 40:845–858

Blanco M, Alberch P 1992 Caenogenesis, developmental variability, and evolution in the carpus and tarsus of the marbled newt, *Triturus marmoratus*. Evolution 46:677–687

Borkhvardt V, Ivashintsova I 1993 Some regularities in the variation of the paw skeleton of Siberian salamander, *Salamandrella keyserlingii* (Amphibia, Hynobiidae). Zool Zh 73:53–67 (in Russian)

Coates M 1995 Limb evolution: fish fins or tetrapod limbs—a simple twist of fate? Curr Biol 5:884–888

Darwin CR 1859 On the origin of species. Murray, London

Duboule D 1994 Temporal colinearity and the phylotypic progression: a basis for the stability of a vertebrate bauplan and the evolution of morphologies through heterochrony. Dev Suppl 135–142

Duellman WE, Trueb L 1986 Biology of amphibians. McGraw–Hill, New York

Gould SJ 1994 Eight little piggies: reflections on natural history. Penguin, London

Gregory RA 1996 Newts and salamanders of Europe. Poyser, London

Hinchliffe JR 1991 Developmental approaches to the problem of transformation of limb structure in evolution. In: Hinchliffe JR, Hurle JM, Summerbell D (eds) Developmental patterning of the vertebrate limb. Plenum Press, New York, p 313–323

Hinchliffe JR 1994 Evolutionary developmental biology of the tetrapod limb. Dev Suppl 163–168

Hinchliffe JR, Griffiths PJ 1983 The prechondrogenic patterns in tetrapod limb development and their phylogenetic significance. In: Goodwin BC, Hodder N, Wylie C (eds) Development and evolution. Cambridge University Press, Cambridge, p 99–121

Holmgren N 1933 On the origin of the tetrapod limb. Acta Zoologica (Stockholm) 14:185–295

Jarvik E 1980 Basic structure and evolution of vertebrates. Vols I & II. Academic Press, New York

Kuzmin SL 1995 The clawed salamanders of Asia. Westarp Wissenschaften, Magdeburg

Marks S, Wake DB, Shubin N 1998 Limb development and evolution in the salamander genera *Desmognathus* and *Bolitoglossa* (Amphibia, Plethodontidae): separating hypotheses of ancestry, function and life history, submitted

Owen R 1849 On the nature of limbs. John van Voorst, London

Schmalhausen II 1915 Development of the extremities in amphibians and their significance for the problem of origin of the extremities of terrestrial vertebrates. Imp Moscow Univ Publ 37:1–263 (in Russian)

Sewertzov A N 1908 Studien uber die entwicklung der extremitaten der heideren Tetrapoda. Bull Soc Natural Moscou (NS) 21:1–432
Shubin N 1991 Implications of the 'Bauplan' for development and evolution of the tetrapod limb. In: Hinchliffe JR, Hurle JM, Summerbell D (eds) Developmental patterning of the vertebrate limb. Plenum Press, New York, p 411–422
Shubin N 1995 The evolution of paired fins and the origin of tetrapod limbs: phylogenetic and transformational approaches. Evol Biol 28:39–86
Shubin N, Alberch P 1986 A morphogenetic approach to the origin and basic organization of the tetrapod limb. Evol Biol 20:319–387
Shubin N, Wake D 1996 Phylogeny, variation and morphological integration. Am Zool 36:51–60
Sitina LA, Medvedeva IM, Godina LB 1987 Development of *Hynobius keyserlingii*. Akademia Nauka, Moscow (in Russian)
Thomson KS 1968 A critical view of the diphyletic theory of rhipidistian-amphibian relationships. In: Ørvig T (ed) Current problems of lower vertebrate phylogeny. Almquist, Stockholm, p 285–306
Vorobyeva EI, Hinchliffe JR 1996a From fins to limbs. Evol Biol 29:263–311
Vorobyeva EI, Hinchliffe JR 1996b Developmental pattern and morphology of *Salamandrella keyserlingii* limbs (Amphibia, Hynobiidae) including some evolutionary aspects. Russ J Herpetol 3:68–81
Vorobyeva EI, Ol'shevskaya OP, Hinchliffe JR 1997 Specific features of development of the paired limbs in *Ranodon sibiricus* Kessler (Hynobiidae, Caudata). Russ J Dev Biol 28:150–158
Wake DB, Hanken J 1996 Direct development in the lungless salamanders: what are the consequences for developmental biology, evolution and phylogenesis? Int J Dev Biol 40:859–869

DISCUSSION

Rudolf Raff: I would like to ask if the hynobiids are the most primitive living tetrapods, or at least ones with complete digits? They are going to be far removed from primitive fossil tetrapods, so how do you know that there isn't simply a lot of heterochronic change going on back and forth within urodeles, and that you have not just fortuitously picked a species near the base of the salamanders. I say this because there is a large phylogenetic gap between living and fossil amphibians.

Hinchliffe: Yes, that's certainly true. There is an enormous gap in the phylogenetic record between temnospondyls and extant urodeles (Carroll 1987), so it could be argued that the similarities in the limb skeleton might be fortuitous. However, on the other hand, the temnospondyl pattern is strikingly similar to that in the primitive urodeles. In particular, the definitive tarsus of the temnospondyls (e.g. *Trematops*; Carroll 1987) is practically identical to that in the primitive hynobiids (e.g. *Salamandrella*) but the similarity is less in advanced urodeles such as *Triturus* (Vorobyeva & Hinchliffe 1996).

Greene: Is anything known about limb development in cryptobranchoids?

Hinchliffe: No, as far as I know, only a relatively small number of species have been studied.

Carroll: You concentrated on tissue condensation and chondrogenesis, but did you compare these with the ossification sequence of the carpal region?

Hinchliffe: We didn't study this in detail, although the carpal region may ossify prior to metamorphosis within the hynobiid group in *Ranodon* compared with later ossification in *Salamandrella*.

Carroll: Rieppel (1992) made specific comparisons among lizards, and he found a close correlation between chondrogenesis and ossification. There are some differences, but as palaeontologists we can only look at ossification. We do have fossils of fairly early amniotes that show the sequence of ossification in the carpal/tarsal region. Therefore, it is possible to check whether the ossification and chondrogenesis sequences were likely to be similar. One should also be concerned about the conflict between the aquatic and terrestrial life stages in amphibians, because one would have to accommodate for the late ossification of the entire carpal/tarsal region in terrestrial forms.

Hinchliffe: Are you suggesting that there may be a link between early chondrogenesis and subsequent ossification such that the latter sequence reflects the former?

Carroll: I am addressing the question of how specifically it would.

Hinchliffe: It's something we could look for. We haven't looked specifically at the ossification pattern in *Ranodon* (where ossification is earlier than in similar stages in *Salamandrella*) but we should. On reflection, chondrogenic rates may well be dissociated from osteogenic rates, given the frequent ossification of digits in advance of the carpus/tarsus. The whole question of whether early condensation also implies rapid chondrogenesis and subsequent ossification needs careful analysis. My hunch, based on studies on the chick leg bud (Hinchliffe & Johnson 1983), is that the timing of the three processes can be dissociated differently in particular elements.

Wagner: You mentioned that there is usually an anteroposterior polarity in the development of the digits. Do you know of any exceptions to this in urodeles?

Hinchliffe: There are some (e.g. *Bolitoglossa subpalmata*) that have a weak polarity, but in general no.

Wagner: Your observations were not within a wider phylogenetic framework, so some of your interpretations depend on what you think is the plesiomorphic stage for the origin of urodeles. Did urodeles arise from a group that had a typical eu-tetrapod pattern of development or did they arise independently of such a group?

Hinchliffe: Within urodeles there is a spectrum that overlaps the amniote pattern, and although previously the emphasis has been on the urodele pattern being fundamentally different, in my opinion this is not so. One can match the urodele limb skeletal bauplan to that of tetrapods. The Shubin & Alberch (1986) branching and segmenting model is fairly mechanistic, so that during condensation a proximal element could segment (as preaxially) or branch (as postaxially), and the definitive structure is thus built in a proximodistal direction. This eu-tetrapod-like pattern fits some urodeles (hynobiids in particular) reasonably well, especially now that the new immunofluorescence methods reveal in these a clear early digital arch like that of eu-tetrapods. However, there are clear examples where this 'fit' is not the case; for example, in *Triturus* distal events sometimes take place earlier than proximal events, for example in the formation of the digit 1–2 and basal commune unit. But this is not the standard pattern for urodeles.

Wagner: Your cell-staining results indicate that development is proximodistal, and it is only in the experiments in which you looked at the secretion of extracellular matrix compounds that you found evidence of distal to proximal development. There is no evidence in the literature that distal parts condense earlier, although they may differentiate earlier. This supports your contention that there is a digital arch, and there's a commonality between urodeles and eu-tetrapods.

Hinchliffe: I agree that we should look at whether there is a close correlation between the timing of matrix synthesis, as revealed by alcian blue staining, and condensation. My impression in *Triturus* is that even at the condensation stage, the distal development of digits 1 and 2, and the basal commune, precedes more proximal events.

Carroll: In terms of ossification, the fossil evidence suggests that the digits and metapodium are formed before the mesopodium.

Wagner: But long-bone ossification always precedes the ossification of nodular bones.

Akam: From the perspective of a developmental geneticist, I would say that by the time you see cells doing something different it is too late. We need markers for stages prior to condensation, because these will tell us when are cells receiving the signals that are building the pattern.

Müller: I would like to ask whether you agree with the interpretation that the basic system of branching and segmenting pre-cartilaginous condensations in the mesenchymal core of growing limb buds is modified throughout evolution by heterochrony, which could be mediated through changes in gene regulation, but that the branching and segmenting system itself is generic to the growth of condensations in a proximodistal direction?

Hinchliffe: Essentially, yes. As I said earlier, the Shubin & Alberch (1986) scheme is mechanistic and depends on the properties of the extracellular matrix being responsible for proximodistal branching. However, it looks as though the more advanced urodeles have diverged from this because there appears to be a physical gap between the proximal and the distal (digit 1–2 and basal commune) positions of these early skeletogenic elements. So this presents a problem for a fairly simple branching/segmenting model. Generally, though, I agree that throughout tetrapod evolution the same branching/segmenting system is frequently modified by heterochrony, as you yourself showed in reptiles and birds (Müller 1991).

Müller: I would argue that the Shubin & Alberch (1986) model still stands if one takes into consideration that *de novo* condensation could occur in an area that is not connected to the primary condensations and where this whole system can start again.

Wagner: But the notion that the basal commune arises from an independent condensation event is weakened by the strong connection to the ulna/fibula.

Wilkins: It seems to me that we do not have any ideas about the genetics which affect local properties, e.g. the size of the condensations and the rates at which they happen. I suspect many of these local differences will be important, but as yet there is a large gap between the information we have on the *Hox* genes, for example, and what we observe in the condensations.

Hinchliffe: It would be fascinating to know which gene regulatory expression domains operate in each precartilage field, but virtually nothing has been published on this, although a start is being made (Shubin et al 1997).

Hall: Some literature is beginning to accumulate on one or two model systems about the molecules involved in these cells condensing together (e.g. neural cell adhesion molecule and fibroblast growth factor), but we are a long way from being able to put that in a comparative context.

Wilkins: I predict that at some point we will find that relatively minor allelic differences cause small variations in the amounts of these molecules and their activities, which will nevertheless have large effects on the sequence of condensations.

Carroll: But we don't have to look at the genetic variation between species. For example, there are dramatic differences in carpal/tarsal number and configuration within populations of small species that are due to the limited space in which the cells are condensed. Within this small space, different carpal/tarsal patterns are formed.

Hinchliffe: Alberch & Gale (1985) have shown that in miniaturized limbs, for example, digital loss occurs, and it is the elements formed last which disappear first. One can get a long way in explaining the diversification of the urodele limbs in this sort of way; for example, in terms of the interaction between the area of available mesenchyme and the skeletal patterning process (Hanken 1985).

Wagner: Limb development in urodeles is fascinating because it has an exceptional amount of variability, whereas limb development in all other tetrapods is strongly canalized. It is an almost unbreakable constraint that the thumb is lost first, and the last digits to be lost are digits 3 and 4 because they are the first to develop. In your adaptive explanation of urodele development, you say that these digits are made early because they are used early, so wouldn't you predict, based on the eu-tetrapod mode of development, that these digits should be digits 3 and 4 if they are derived from eu-tetrapods.

Hinchliffe: It is worth proposing that the precocious development in many urodeles of digits 1 and 2 is a locomotory adaptation of free-living larvae because the limb in direct-developing larvae appears much more amniote like in its development, even in advanced species such as *B. subpalmata* (Marks et al 1998). To be speculative, there seems to be a basic pattern that is pulled over by natural selection in the direction of earlier anterior digit development, especially in the most advanced salamanders, but as soon as you get a direct-developing form it becomes more amniote like. This loss of the free-living larva has happened several times, so in principle we can check this idea in direct-developing species from other groups. So if this hypothesis is accepted, it does not seem necessary to suppose digit loss anteriorly and neomorphic appearance of new digits posteriorly to explain the unusual pattern of urodele limb development.

References

Alberch P, Gale E 1985 A developmental analysis of an evolutionary trend: digital reduction in amphibians. Evolution 39:8–23

Carroll RL 1987 Vertebrate paleontology and evolution. WH Freeman, New York

Hanken J 1985 Morphological novelty in the limb skeleton accompanies miniaturization in salamanders. Science 229:871–874

Hinchliffe JR, Johnson DR 1983 Growth of cartilage. In: Hall BK (ed) Cartilage, vol 2. Academic Press, New York, p 255–294

Marks S, Wake DB, Shubin N 1998 Limb development and evolution in the salamander genera *Desmognathus* and *Bolitoglossa* (Amphibia, Plethodontidae): separating hypotheses of ancestry, function and life history, submitted

Müller GB 1991 Evolutionary transformation of limb pattern: heterochrony and secondary fusion. In: Hinchliffe JR, Hurle JM, Summerbell D (eds) Developmental patterning of the vertebrate limb. Plenum Press, New York, p 395–405

Rieppel O 1992 Studies on skeleton formation in reptiles. I. The postembryonic development of the skeleton in *Cyrtodactylus pubisulus* (Reptilia: Gekkonidae). J Zool (Lond) 227:87–100

Shubin N, Alberch P 1986 A morphogenetic approach to the origin and basic organization of the tetrapod limb. Evol Biol 20:319–387

Shubin N, Tabin C, Carroll RL 1997 Fossils, genes and the evolution of animal limbs. Nature 388:639–648

Vorobyeva EI, Hinchliffe JR 1996 Developmental pattern and morphology of *Salamandrella keyserlingii* limbs (Amphibia, Hynobiidae) including some evolutionary aspects. Russ J Herpetol 3:68–81

Larval homologies and radical evolutionary changes in early development

Rudolf A. Raff

Indiana Molecular Biology Institute, and Department of Biology, Indiana University, Bloomington, IN 47405, USA

Abstract. Larval forms are highly conserved in evolution, and phylogeneticists have used shared larval features to link disparate phyla. Despite long-term conservation, early development has in some cases evolved radically. Analysis of evolutionary change depends on identification of homologues, and this concept of descent with modification applies to embryo cells and territories as well. Difficulties arise because evolutionary changes in development can obscure homologies. Even more difficult, threshold effects can yield changes in process whereby apparently homologous features can arise from new precursors or pathways. We have observed phenomena of this type in closely related sea urchins that differ in developmental mode. A species developing via a complex feeding larva and its congener, which develops directly, have different embryonic cell lineages and divergent patterns of early development, but converge on the adult sea urchin body plan. Despite differences in embryonic developmental pathways, conserved gene expression territories are evident, as are territories whose homologies are in doubt. The highly derived development of the direct developer evidently arises from an interplay of novel organization of the egg, loss of expression of regulatory genes involved in production of feeding larval features, and changes in site and timing of expression of a number of genes.

1999 Homology. Wiley, Chichester (Novartis Foundation Symposium 222) p 110–124

Development and homology

In 1843 Richard Owen defined homology as 'the same organ in different animals under every variety of form and function'. Owen's non-evolutionary concept was transformed by Darwin into the powerful idea that homologues arise from descent with modification. As Van Valen (1982) put it, homologues represent an evolutionary continuity of information. The classical criteria for homologues are position in a comparable system of features; having similar compositional characteristics; and a historical transition (generally from the fossil record) between ancestral and descendant structure.

Phylogeneticists assume that natural processes, such as evolution, operate in a parsimonious way, with evolution following the line of fewest changes. Indeed, more related forms are likely to be more similar than more distant ones, and homologies in development are most easily demonstrated among similar forms. Continuity of information between generations of organisms is manifested through development. Homologous features in two related organisms should arise by similar developmental processes, and homology should be most readily recognized by ontogenetic criteria. Many cases of homologies have been so recognized. However, homologues are not always easy to unequivocally identify because a variety of processes can produce similarities that masquerade as homologies. And, some processes can change homologues in ways that obscure recognition.

Most intriguingly, features that we regard as homologous from morphological and phylogenetic criteria can arise in different ways in development (Rieppel 1994, Shubin 1994). As first pointed out by von Baer in the 1820s, animals within a phylum, such as vertebrates, share a common body plan, and in their development share a phylotypic stage in which the body plan elements characteristic of the phylum appear. The process of early development from the egg to the phylotypic stage should be at least as conserved as the pattern of the phylotypic stage. One might reasonably expect mechanisms of early development to be especially resistant to modification because all subsequent development derives from early processes. Traditionally, features of early development and conserved larval stages, even between phyla, have been regarded as strong homologous characters for the inference of phylogeny. The division of animals into protostome and deuterostome superphyla is based on the ideas that embryonic similarities are homologous and have been largely immutable over hundreds of millions of years (Raff 1996).

A view of development from an evolutionary perspective is both more confounding and more interesting. Early development is highly evolvable, even among closely related species. The evolutionary portrait of ontogeny may be that of an hourglass, as shown in Fig. 1. In this diagram, embryos of two related species follow different early developmental trajectories, but converge on a similar phylotypic stage. It is important to note that the phylotypic stages of related organisms bear major features in common, but also have evolved significantly (Richardson et al 1998). Divergence in post-phylotypic developmental trajectories yields variant adult species morphologies, as suggested by von Baer. The divergence of pre-phylotypic stage pathways can be extreme. For example, polyembryonic parasitic wasps have a bizarre early developmental pathway that does not resemble typical insect early development. Polyembryonic development produces 2000 embryos from one egg. Nonetheless, these secondary embryos develop via a characteristic insect phylotypic stage (Grbic et al 1998).

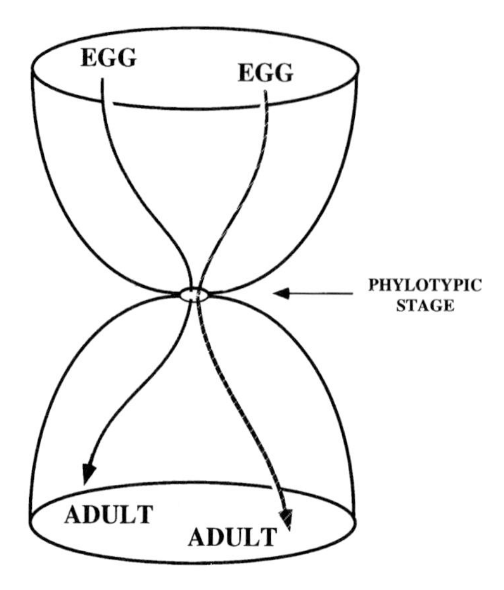

FIG. 1. The developmental hourglass. The developmental trajectories of two species can differ both in early and late development, with conservation of the phylotypic stage (the constriction of the hourglass). After Raff (1996).

Evolution of developmental mode

Most animal phyla exhibit complex life histories, in which a larva distinct from the adult is produced, then metamorphosis to the adult takes place. This pattern includes sea urchins, which typically develop via a pluteus larva, a bilaterally symmetrical feeding larva. Although this is the primitive mode of echinoid development, retained for at least 250 million years, about 20% of sea urchin species have independently evolved direct development from the ancestral mode (Raff 1996, Wray 1996). Analogous shifts have occurred among numerous taxa, including frogs, starfish, ascidians and molluscs. Radical changes have occurred independently among congeneric species (Hart et al 1997, Hadfield et al 1995, McMillan et al 1992), which makes these examples suitable for the experimental study of evolution of early development.

The Australian sea urchin genus *Heliocidaris* contains two species. *Heliocidaris tuberculata*, which has a pluteus larva, and *Heliocidaris erythrogramma*, which develops directly into a juvenile adult (Fig. 2). We have sought homologues among larval features of the two in order to understand the transformation between developmental modes. Three hierarchical levels of homology can be hypothesized for embryos: morphological features, cell lineages and gene expression. *H. erythrogramma* is greatly simplified morphologically, even though apparently homologous morphological larval features are present. These include the archenteron, coelom, vestibule, skeletogenic mesenchyme and pigment cells. However, evolutionary transformations of morphology have affected our understanding of homologous features. First, apparently homologous larval features are distinct in both morphogenesis and in origins. Thus, the archenteron arises through a different set of cellular movements in *H. erythrogramma* than in the pluteus (Wray & Raff 1991), and it does not form a feeding larval gut. Second, apparently homologous features can exhibit different behaviours. The skeletogenic mesenchyme cells of *H. tuberculata* secrete long simple larval skeletal rods, but in *H. erythrogramma* onset of skeletogenesis is delayed and consists primarily of adult skeletal meshwork (stereom) elements. However, skeletogenic cells exhibit conserved expression of at least some characteristic genes (Parks et al 1988). Third, features can be lost altogether, which changes topological relationships. The oral ectoderm is lacking altogether in *H. erythrogramma* (Raff et al 1999). Homology of morphological features is inferred on the basis of classic criteria of composition and location within the morphological system. However, the compositional criterion becomes iffy at the gene expression level.

Cell lineages and the origins of homologous features

Simply looking at the changed morphology of the *H. erythrogramma* larva is insufficient to determine the origins or movements of presumably homologous cells. This has to be done by defining the lineages of embryonic cells. Maps and lineages for a 32-cell indirect-developing sea urchin and 32-cell *H. erythrogramma* are shown in Fig. 3. Among the important changes are cleavage pattern, timing of allocation of cell fates and origins of lineage founder cells (Wray & Raff 1990). In indirect developers, the micromeres produced at fourth cleavage yield a skeletogenic founder cell, and a coelomic founder at the next cleavage. Both cell types are present in *H. erythrogramma*, but they do not arise from a micromere founder. There are also substantial differences in cell allocation. In the indirect developer, vestibular ectoderm arises from some daughters of two mesomeres. In *H. erythrogramma*, a fourth of the cells of the 32-cell embryo give rise to vestibular ectoderm. These changes pose a quandary if a developmental criterion is to be taken seriously, and there are problems with other criteria as well. For example, the criterion of localization of a structure within a field of structures is violated by the *H. erythrogramma* vestibular ectoderm. In indirect developers, the embryonic ectoderm consists of both oral and aboral ectoderm. In development of the pluteus, the vestibule forms as part of the oral ectoderm, but this topological relationship has been lost in *H. erythrogramma*.

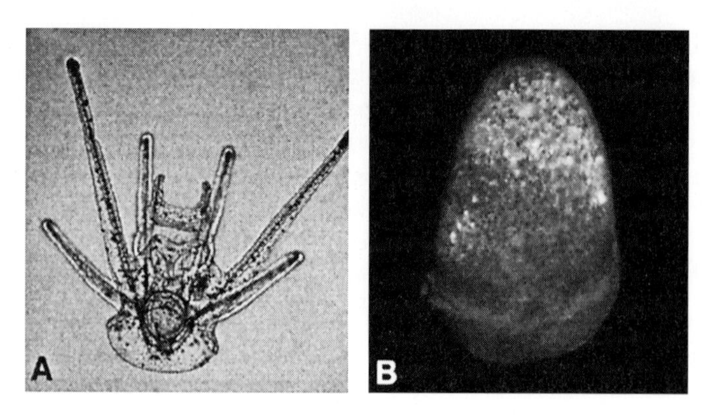

FIG. 2. (A) The pluteus larva of *Heliocidaris tuberculata*. (B) The highly modified larva of *Heliocidaris erythrogramma*. Courtesy of E. Popodi.

Gene expression patterns and homology inferences

The transitional criterion is classically used for evolutionary transitions as traced in the fossil record, which is not likely to be informative with most embryos. One might propose a developmental version of the transitional criterion based on cellular origins, and that has been extensively used in defining similarities among spiralian embryos (Anderson 1973). However, *H. erythrogramma* shows us that cell lineages can evolve to the point where a developmental transitional criterion based on cell lineage cannot be readily applied. In the case of *Heliocidaris*, comparisons are being made among closely related species, but difficulties may be proportionately more serious for desired comparisons at greater phylogenetic distances. Where sought between animals that have different body plans, developmental homologies become even more difficult to pin down. Examples include such questions as whether there are homologues of the notochord in echinoderms and hemichordates, or if segmentation is homologous between annelids and arthropods.

The application of the last classic criterion, the criterion of compositional similarity, has recently become popular in seeking answers to homology questions. The compositional similarities that are sought are those in expression of regulatory genes involved in patterning the structure in question. This approach has become possible because major regulatory genes are shared among all animal groups.

The use of gene expression as a criterion for inferring homologies has reopened a long tradition in zoology, the possibility of elucidating underlying homologies and even transformations between dissimilar body plans, a favourite activity in the late 19th and early 20th centuries. Gene expression patterns certainly suggest

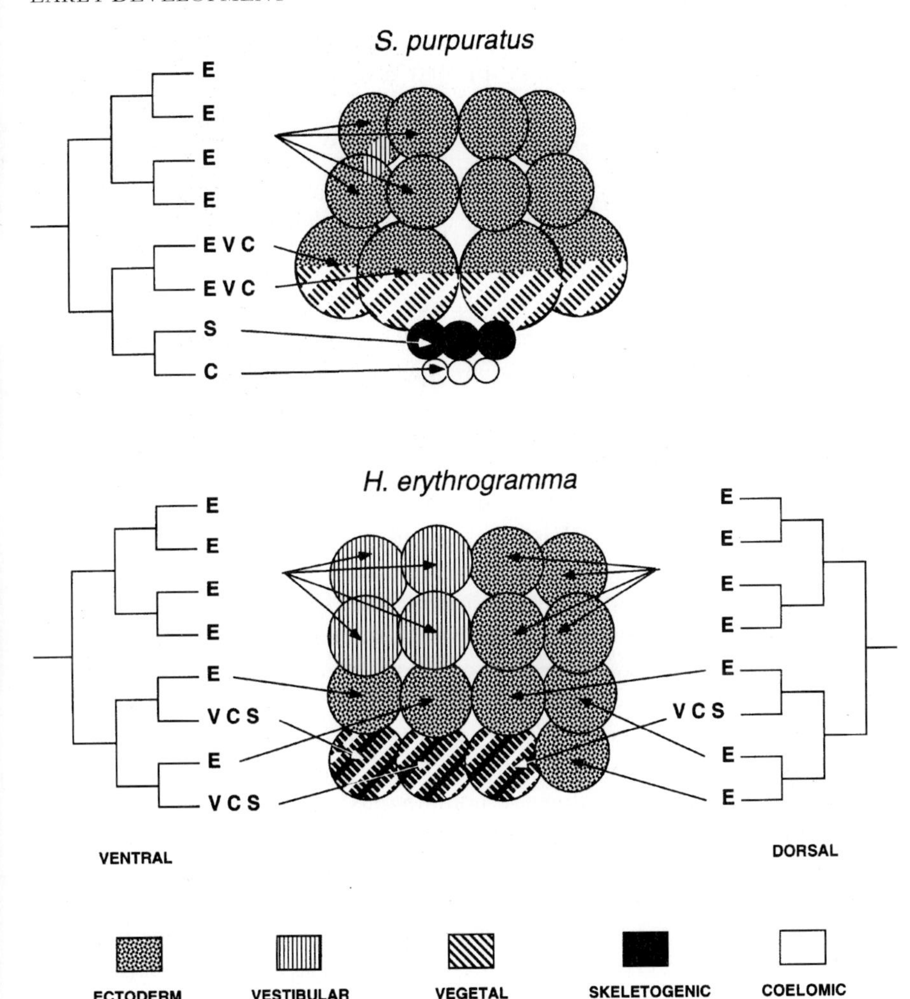

FIG. 3. Fate maps of an indirect-developing sea urchin (*Strongylocentrotus purpuratus*) and *Heliocidaris erythrogramma*. Cell lineage diagrams are shown for a quarter of the *S. purpuratus* embryo, and for quarters lying in the distinctly different ventral and dorsal halves of *H. erythrogramma*. C, coelom; E, ectoderm; S, skeletogenic cells; V, vegetal plate-derived cells. Cell fates are pattern coded. After Wray & Raff (1990).

some interesting deep homologies. For example, patterns of expression of genes involved in establishment of dorsoventral body axes in arthropods and vertebrates suggest that the classical anatomical interpretation that these animals are upside-down equivalents has some foundation (De Robertis & Sasai 1996). Other intriguing patterns of gene expression have also emerged. For example, the gene *distalless* is expressed in developing projections/appendages of a number of phyla (Panganiban et al 1997). *distalless* is expressed in annelid parapods, vertebrate fins and limbs, arthropod appendages, echinoderm tube feet, and ascidian siphons and ampullae. Are all these structures somehow homologous (although not necessarily as limbs), or has *distalless* been co-opted to serve an analogous function in projections from the body axis?

Comparisons of presumptive molecular patterning homologies between closely related and similar forms, such as the fruitfly *Drosophila* and the housefly *Musca* (Sommer & Tautz 1991), reveal similar genes expressed in similar patterns, a parsimonious outcome. We can ask how well gene expression patterns do as indicators of homology among related animals in which there are substantial modifications in development, as in *H. erythrogramma*. The answer varies by gene. The highly conserved and intricately regulated actin gene family shows a great conservation of genes and tissue-type expression patterns among vertebrates, but has varied rapidly among sea urchins, including among indirect developers that all produce the same larval form (Kissinger & Raff 1998). On the other hand, Wnt5, an intracellular signalling protein, is highly conserved in expression between indirect-developing sea urchin embryos and the direct-developing *H. erythrogramma*, although timing of expression is shifted (Ferkowicz & Raff 1999). Among echinoderm classes, homeobox-containing transcription factor genes have undergone substantial co-option, making them unreliable for deciding homologies among echinoderms (Lowe & Wray 1997).

Co-option of genes and genetic pathways expressed in new locations can lead to homologous processes producing non-homologous structures (Gilbert et al 1996), and can be a major contributor to perceived homology difficulties. Our observations of gene expression patterns in *H. erythrogramma* gene expression territories suggest that co-option of gene expression is prevalent and contributes to the evolution of novel larval features. The molecular criteria for inferring homologies become blurred as a result. Both indirect-developing sea urchins and *H. erythrogramma* are composed of gene expression territories. Figure 4 compares the gene expression territories of an indirect-developing sea urchin (Coffman & Davidson 1992) and *H. erythrogramma* (Raff 1999). Several territories appear to be homologous, if not by a criterion of developmental origin, at least by cell type and gene expression pattern. Ectodermal territories are more difficult to match.

The primary ectodermal territories of an indirect developer are oral and aboral (Coffman & Davidson 1992). Vestibular ectoderm arises later within the oral territory, but in *H. erythrogramma*, there is no oral ectodermal territory. Although the ciliary band forms a boundary between the oral and aboral territories in the pluteus, it forms no territory boundary in *H. erythrogramma*, and so the vestibular ectoderm abuts extravestibular ectoderm. *H. erythrogramma* extravestibular ectoderm has columnar cells rather than squamous as in the pluteus. It is also difficult to homologize by pattern of gene expression. Extravestibular ectoderm does not express either CyIII actin or arylsulfatase, characteristic gene markers of aboral ectoderm in the pluteus (Kissinger & Raff 1998, Haag & Raff 1998). Some gene expression features of vestibular ectoderm have also changed. In *H. erythrogramma*, arylsulfatase is expressed in the vestibular ectoderm, but it is not in the pluteus (Haag & Raff 1998). However, there is little problem in homologizing vestibular ectoderms because of their shared complex morphogenetic pathways in building the juvenile sea urchin in both *H. erythrogramma* and indirect developers.

Threshold effects and the evolution of novelty in development

The idea that homology arises from a continuity of information in evolution suggests a kind of gradualistic view in which phenotypic change maps in a linear manner on genomic change. Although this may be correct for the evolution of many homologues, there also appear to be mechanisms that operate to produce evolutionary changes in development that exhibit non-linear threshold behaviour. How might this happen? The first possibility is exemplified by a particularly fascinating idea suggested for tetrapod limbs by Goodwin & Trainor (1983). The limb arises from a developmental field that is governed by global transformational rules. Evolutionary changes in patterning will thus result from changes in the parameters of the field. The consequence of this kind of mechanism is that individual elements (say digits) do not have a discrete homology because they do not have unique developmental identities. Appearance of a novel element would not necessarily arise through transformation of any ancestral element, but arise out of changes in the global properties of the field. In evolution the fields of related taxa would represent developmental homologues, not individual elements.

A second way in which development can be transformed is through changes in expression of high level genetic hierarchies to produce novel features within an embryo. There is a phenomenon widely observed in embryos that has been classically called 'regulation' by embryologists. Many embryos can restore missing features resulting from perturbation. A well-known example is the ability of a sea urchin embryo separated at the first cleavage into two half embryos. Both halves can develop into normal embryos, and even undergo metamorphosis. A more subtle example is the ability of sea urchin embryos, from which the precursor cells (the micromeres) of the skeleton secreting cells are removed, to generate skeletogenic cells from a cell lineage that normally does not produce this fate (Ettensohn & McClay 1988). The conversion involves cells from the vegetal plate gene expression territory switching territory identity, a kind of homoeosis. We generally observe embryonic regulation effects by experimental

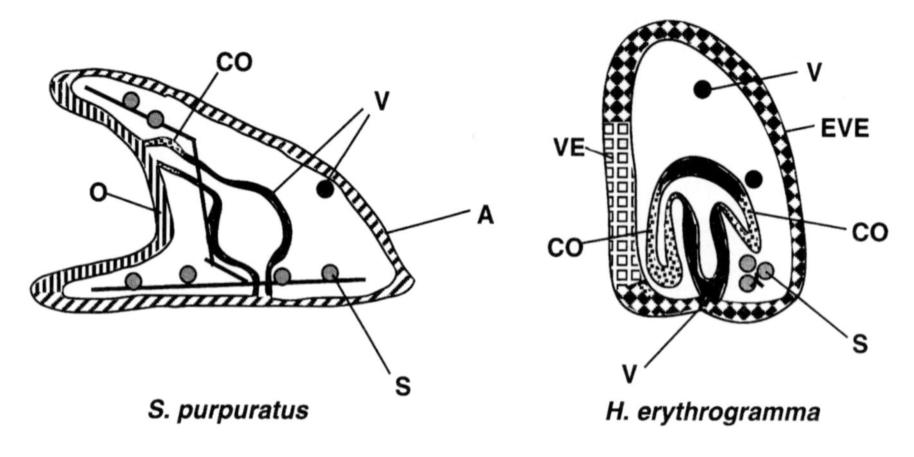

FIG. 4. Gene expression territories of an indirect-developing sea urchin (*Strongylocentrotus purpuratus*) and *Heliocidaris erythrogramma*. A, aboral ectoderm; CO, coelom; EVE, extravestibular ectoderm; O, oral ectoderm; S, skeletogenic cells; V, vegetal plate-derived cells; VE, vestibular ectoderm. *S. purpuratus* after Coffman & Davidson (1992).

perturbation of embryos, but mutational changes might lead to analogous evolutionary changes that can form the basis of a modified ontogeny and exhibit reorganized pathways to 'homologous' embryonic and larval features. The evolutionary changes in cell lineage observed in *H. erythrogramma* appear to have arisen in this way, reflecting flexible developmental regulatory properties of indirect-developing echinoids (Raff 1996).

A third way in which non-linear processes can affect the development of homologues is through co-option (Raff 1996). Entire regulatory gene systems or fields can be co-opted to be expressed in a new location within the embryo. The result is appearance of a molecular serial homologue. The *Hox* gene expression pattern of vertebrate appendages suggests that these morphological serial homologues might have originated in this way (Tabin & Laufer 1993).

Finally, although high level regulatory genes are conserved, and may retain conserved patterns of expression, they can get incorporated into new gene circuitries and new downstream cascades (Nagy 1998). This can lead to 'homologous' features arising via different genetic foundations. The anterior (head) ends of insects are surely homologous, yet the determination of the head end of flies is based on a gene cascade starting with, and dependent on, *bicoid*, a gene that exists only within the Diptera (Nagy 1998). Changes in regulatory gene function are visible in at least two ways in the evolution of *H. erythrogramma*. The histospecific genes that characterize sea urchin gene expression territories are under complex regulatory controls that interact on their promoters (Kirchhamer et al

1996). Their patterns of expression are changed in *H. erythrogramma* (Kissinger & Raff 1998) and in one case clearly by changes in trans-acting factors (Klueg et al 1997). Furthermore, cross-species hybrid experiments reveal the existence of dominant regulatory factors lost in *H. erythrogramma* (Raff et al 1999).

Conclusions

Returning to Owen's definition of homology as the 'same organ in different animals under every variety of form and function', we find that development can be confounding as well as helpful as to what constitutes the 'same'. With respect to Van Valen's view that homologues represent a continuity of information, we find that some of the processes that underlie the evolution of development can confound what we mean by 'continuity' as well. Should all this be a cause for despair? No. Biologists have historically used homologies to trace evolutionary histories and phylogenetic relationships. A deeper understanding reveals that this cannot be a tidy programme. That is on the face of it unfortunate, and has generated much hand wringing on the usefulness of the homology concept. However, where developmental homologies are difficult to identify because of process shifts in ontogeny, we are actually being told interesting things about evolution. Ambiguities in development of homologues in embryos reveal where and how evolutionary changes occur and thus, although confounding, are difficult Rosetta Stones needed to understand how evolutionary novelties arise.

Acknowledgements

I thank the National Institutes of Health and the National Science Foundation for support, and the School of Biological Sciences, University of Sydney, Australia for the use of its facilities in this work.

References

Anderson DT 1973 Embryology and phylogeny in annelids and arthropods. Pergamon Press, Oxford

Bolker JA, Raff RA 1996 Developmental genetics and traditional homology. BioEssays 18:489–494

Coffman JA, Davidson EH 1992 Expression of spatially regulated genes in the sea urchin embryo. Curr Opin Genet Dev 2:260–268

De Robertis EM, Sasai Y 1996 A common plan for dorsoventral patterning in Bilateria. Nature 380:37–40

Ettensohn CA, McClay DR 1988 Cell lineage conversion in the sea urchin embryo. Dev Biol 125:396–409

Ferkowicz MJ, Raff RA 1999 Patterns of *wnt-5* expression in early sea urchin development: heterochronies associated with developmental mode, in prep

Gilbert SF, Opitz JM, Raff RA 1996 Resynthesizing evolutionary and developmental biology. Dev Biol 173:357–372

Goodwin BC, Trainor C 1983 The ontogeny and phylogeny of the pentadactyl limb. In Goodwin BC, Holder N, Wylie CC (eds) Development and evolution. Cambridge University Press, Cambridge, p 75–98

Grbic M, Nagy LM, Strand MR 1998 Development of polyembryonic insects: a major departure from typical insect embryogenesis. Dev Genes Evol 208:69–81

Haag EH, Raff RA 1998 Isolation and characterization of three mRNAs enriched in embryos of the direct-developing sea urchin *Heliocidaris erythrogramma*: evolution of larval ectoderm. Dev Genes Evol 208:188–204

Hadfield KA, Swalla BJ, Jeffery WR 1995 Multiple origins of anural development in ascidians inferred from rDNA sequences. J Mol Evol 40:413–427

Hart MW, Byrne MM, Smith MJ 1997 Molecular phylogenetic analysis of life history evolution in asterinid starfish. Evolution 51:1848–1861

Kirchhamer CV, Yuh CH, Davidson EH 1996 Modular cis-regulatory organization of developmentally expressed genes: two genes transcribed territorially in the sea urchin embryo, and additional examples. Proc Natl Acad Sci USA 93:9322–9328

Kissinger JC Raff RA 1998 Evolutionary changes in sites and timing of expression of actin genes in embryos of the direct- and indirect-developing sea urchins *Heliocidaris erythrogramma* and *H. tuberculata*. Dev Genes Evol 208:82–93

Klueg K, Harkey MA, Raff RA 1997 Mechanisms of evolutionary changes in timing, spatial expression, and mRNA processing in the *msp130* gene in a direct-developing sea urchin, *Heliocidaris erythrogramma*. Dev Biol 182:121–133

Lowe CJ, Wray GA 1997 Radical alterations in the roles of homeobox genes during echinoderm evolution. Nature 389:718–721

McMillan WO, Raff RA, Palumbi SR 1992 Population genetic consequences of developmental evolution in sea urchins (Genus *Heliocidaris*). Evolution 46:1299–1312

Nagy L 1998 Changing patterns of gene regulation in the evolution of arthropod morphology. Am Zool, in press

Owen R 1843 Lectures on the comparative anatomy and physiology of the invertebrate animals. Longman Brown Green & Longmans, London

Panganiban G, Irvine SM, Lowe C et al 1997 The origin and evolution of animal appendages. Proc Natl Acad Sci USA 94:5162–5166

Parks AL, Parr BA, Chin JE, Leaf DS, Raff RA 1988 Molecular analysis of heterochrony in the evolution of direct development in sea urchins. J Evol Biol 1:27–44

Raff EC, Popodi EM, Sly B, Turner FR, Villinski JT, Raff RA 1999 A novel ontogenetic pathway in hybrid embryos between species with different modes of development, submitted

Raff RA 1996 The shape of life: genes, development and the evolution of animal form. University of Chicago Press, Chicago, IL

Raff RA 1999 Modularity and dissociation in the evolution of development: evolution of sea urchin embryo gene expression territories, in prep

Richardson MK, Allen SP, Wright GM, Raynaud A, Hanken J 1998 Somite number and vertebrate evolution. Development 125:151–160

Rieppel O 1994 Homology, topology, and typology: the history of modern debates. In: Hall BK (ed) Homology: the hierarchical basis of comparative biology. Academic Press, San Diego, CA, p 63–100

Shubin NH 1994 History, ontogeny, and evolution of the archetype. In: Hall BK (ed) Homology: the hierarchical basis of comparative biology. Academic Press, San Diego, CA, p 249–271

Sommer RJ, Tautz D 1991 Segmentation gene expression in the housefly *Musca domestica*. Development 113:419–430

Tabin C, Laufer E 1993 *Hox* genes and serial homology. Nature 361:692–693

Van Valen L 1982 Homology and causes. J Morphol 173:305–312

Wray GA 1996 Parallel evolution of nonfeeding larvae in echinoids. Systematic Biology 45:308–322
Wray GA, Raff RA 1990 Novel origins of lineage founder cells in the direct-developing sea urchin *Heliocidaris erythrogramma*. Dev Biol 141:41–54
Wray GA, Raff RA 1991 Rapid evolution of gastrulation mechanisms in a direct-developing sea urchin. Evolution 45:1741–1750

DISCUSSION

Wray: It is worth remembering that there is a species of sea urchin which has a larva that is not a hybrid but has many of these properties (*Phyllacanthus imperialis*). The larva of this species is not unlike the synthetic one you made. Its cell lineages are probably similarly modified because it has a similar, but independently derived, cleavage pattern. It is still making many of the cell types that are present in a feeding larva, yet it doesn't have a feeding larva. It also seems to have an intermediate morphology. Therefore, what you're looking at may not only be experimentally tractable, but may have also happened in a real situation.

Rudolf Raff: Many, although not all, of these effects may be heterochronic, so I don't know the dynamics of making a reduced pluteus of something like *Phyllacanthus* from an ancestor with a complete feeding pluteus.

Wray: But it has that same difference in cleavage pattern, so there may be an interesting parallel.

Lacalli: Rudolf Raff mentioned the idea of latent homologies, which I find interesting. Thinking of my own work on echinoderm larvae, I'm reminded of the time I mixed some starfish bipinnaria larvae in with pluteus larvae in order to photograph them together. I realized then how tiny, specialized and advanced the pluteus larvae were compared with the large, floppy and rather slow-moving starfish larvae. My view is that the ancestral larva was probably much more like the bipinnaria than the pluteus, though not all echinoderm specialists would agree. It's nevertheless interesting that by scrambling the genes, some of which presumably were involved in driving pluteus evolution towards specialization, you seem to have uncovered an underlying and more fundamental morphology.

Rudolf Raff: This may or may not be the case. It intrigues me that architectonic rules may be present, and that the conserved epigenetic interactions between ectodermal territories give rise to the conserved aspects of echinoderm larvae. It may be that we are doing nothing more in cross-species hybrids between a direct- and an indirect-developing sea urchin than creating this conserved set of territories, and they then assemble a generalized echinoderm larval form. This is a reasonable alternative to the idea of latent homology at the gene level.

Lacalli: Other terms spring to mind, such as recovery or resurrection!

Hall: Are you uncovering elements of the phylotypic stage in this group?

Rudolf Raff: I don't think so. I would say that the phylotypic stage for echinoderms lies in the pentameral rudiment. The rudiment of *Heliocidaris erythrogramma* is identical to that of indirect-developing sea urchins.

Lacalli: Not all echinoderm biologists would agree with that. Some, including myself, would argue that the phylotypic stage occurs in the larva and not in the adult.

Maynard Smith: But if I understand this correctly, the thing that actually develops into the sea urchin is not the pluteus larva but a particular rudiment of the pluteus larva. If you try to recognize similarities of form or gene activity between *H. erythrogramma* development and the development of that rudiment from other species of echinoderm do you then see morphological similarities?

Rudolf Raff: We compared them using confocal microscopy and they are morphologically identical, although we do see that some different genes are expressed.

Wilkins: You said that in *H. erythrogramma* the ciliary band doesn't mark a boundary. I would say that it must mark a boundary of some kind because it forms in a particular place. This seems to highlight the problems in this field when talking about homologies. We see 'hard' characters in certain kinds of species, such as oral and aboral ectoderm, that are separated by a ciliary band. We then look at related systems or experimentally change things and don't see those hard characters, but rather other things that are more difficult to interpret, and we begin to feel lost. I would maintain that when we look at the hard characters we assign a certain fundamental property to them and then try to read it back into the underlying genetical basis. Frequently, a change in just one or two gene activities will prevent the development of some of those traits. What is your opinion of this?

Rudolf Raff: If I went back to the laboratory tomorrow and looked at gene 26, for example, and found a patch underneath the ciliary band of *H. erythrogramma*, I would say that I had found a territory boundary. The fact that we haven't yet found a territory boundary doesn't mean that there isn't one, although morphologically the cells above and below are similar, so at this point we don't think a boundary is there. The morphology of the hybrid, in which the ciliary band separates two different fields, oral from aboral ectoderm, suggests that these tissues are present; a ciliary band, perhaps specified epigenetically, masks a boundary.

Wilkins: I'm not questioning whether or not those traits are real. I'm questioning the significance being placed on them. There is a tendency to reify certain traits, and to work with them and feel comfortable with them. But if we make small genetic changes and lose those traits, we then start wondering whether they are homologous or not, and this is the wrong question.

Wray: What's the right question?

Wilkins: What genetic machinery makes those traits?

Akam: But the trait is not the characteristic of interest. The trait is the outcome of the process, and so it is the changes in the process that we need to understand.

Lacalli: There may be a simple answer to why the ciliary bands in these organisms act as boundaries in some instances but not others. A related larval type, the tornaria of hemichordates, has two ciliary bands; a circumoral feeding band at the oral/aboral boundary, and a separate telotroch around the anus that is not part of the oral/aboral system. Similar posterior bands and band fragments can be identified in some cases in echinoderm larvae, so the presence of two such systems, or at least the potential to form them, could be part of the ancestral plan.

Rudolf Raff: That makes sense, and it is likely here. Indirect-developing sea urchin larvae have two kinds of ciliary bands. One is a boundary, and the other is called the epaulet, which lies below the anus. The ciliary band of *H. erythrogramma* may well be homologous to epaulets.

Lacalli: I still think parts of an older plan could be recovered, the telotroch for example, and be assembled to make something that looks new.

Wray: This is unlikely—otherwise we would have to assume that species acquired the trait at least 400 million years after it was lost. As Rudolf Raff pointed out, the hybrid only has one band, it's just that it is a different shape, so it makes more sense to consider that its position has just shifted around.

Lacalli: I wonder about this. In our work on crinoid embryogenesis we find indirect evidence for the participation of two organizing centres in band formation, one centred at the mouth and one at the anus. I can easily imagine the bands themselves being lost and recovered during evolution so long as the organizing centres remain intact.

Wray: That's a hypothetical situation. We don't know that there is an organizing centre there, and we don't know whether there is a pattern-forming system that has been maintained but not utilized.

Lacalli: My feeling that the circum-anal part of the ciliary band system is distinct from the oral field is based mainly on observations on crinoids (Lacalli & West 1986): the circum-anal field evidently forms earlier because the ciliary band cells line up sooner and with greater precision than they do around the oral field.

Wray: That doesn't seem to be in the case for sea cucumbers.

Roth: In terms of cellular interactions and territories, could anything additional be learned from studying chimerae instead of hybrids?

Rudolf Raff: We're doing those experiments. Sharon Murisuk in my laboratory has devised ways of taking *H. erythrogramma* apart and putting it back together. She showed that, for example, the ectoderm of the archenteron only forms a vesicle, but if she puts the archenteron back a full suite of structures develop. We are now going to make chimerae between the hybrid and *H. erythrogramma* to look at interactions between cell types.

Hall: You could also combine separate cell lineages.

Rudolf Raff: Yes, although we will first have to determine the cell lineages of the hybrids.

Abouheif: I'm always impressed at how many changes occur in early development of echinoderms and at their variability. I just wondered how unique this is to echinoderms, given that they are such a bizarre phylum and they have a weird life history. Could these results be extrapolated to the rest of the animal kingdom?

My second question is, given that various actin genes play different roles at different times in development, is it safe to use these genes to mark different embryological territories?

Rudolf Raff: When biologists work on development they pick a particular set of organisms to look at, and those become the ones that are reified in the textbooks as the way 'sea urchins' or 'frogs' develop, for example. It is only when you take a look at the exceptions in nature you find exactly how much variation is present.

In answer to your second question, statements about whether parts of an embryo are homologous are often based on whether certain genetic markers are expressed in particular positions. It is clear from our studies and those of others that this isn't a safe way to try to establish homologies because of co-option of gene expression patterns. It is possible that the underlying regulatory gene expression patterns are not so variable, but the downstream genes that execute patterns of development clearly can vary radically. There certainly are differences in behaviour among higher-level taxa. For example, actins are highly conserved in vertebrates, but vary in family membership and expression pattern over a million years among sea urchins.

Wray: There are few animal phyla that don't contain at least this range of diversity in early development.

Reference

Lacalli TC, West JE 1986 Ciliary band formation in the doliolaria larva of *Florometra*. I. The development of normal epithelial pattern. J Embryol Exp Morphol 96:303–323

A research programme for testing the biological homology concept

Günter P. Wagner

Department of Ecology and Evolutionary Biology, Yale University, New Haven, CT 06520-8106, USA

Abstract. The classical homology concept has served as a heuristic principle for organizing the enormous wealth of information on comparative anatomical patterns across a wide range of organisms. However, the classical homology concept reaches its limit as knowledge of the evolutionary, genetic and developmental processes that underlie these anatomical patterns increases. The biological homology concept places the known anatomical patterns into a mechanistic context and asserts that character identity is based on common variational properties. In this chapter a research programme for testing the biological homology concept that involves the following steps is outlined: (1) identifying of two or more putative homologues in a clade; (2) determining the phylogenetic distribution of the putative homologues; (3) describing the intra- and interspecific variation patterns of each putative homologue; (4) describing the development of each putative homologue, and determining if modes of development and distribution of homologues are phylogenetically congruent; and (5) providing and testing a model of how differences in modes of development between putative homologues effect differences in variational tendencies. The goal is to demonstrate a link between developmental and variational differences of two homologues.

1999 Homology. Wiley, Chichester (Novartis Foundation Symposium 222) p 125–140

No two characters are literally identical. To call them nevertheless 'the same' by giving them the same name, like tetrapod limb or neocortex, makes an implicit mechanistic assumption: it is assumed that the differences between instances of a homologue (e.g. the bird wing and a mouse hand) are more likely to occur by natural variation than the differences between different homologues, like the differences between an eye and a limb. This argument leads to the basic idea of the biological homology concept, namely that character identity (i.e. homology) is a hypothesis about the variational properties of characters (constraints and degrees of freedom).

In this chapter I first summarize the basic assumptions and implications of the biological homology concept, and then provide an outline of a research programme to test the biological homology concept.

The core assumptions of the biological homology concept

The biological homology concept has been discussed from a variety of viewpoints (Donoghue 1992, Osche 1973, Roth 1984, 1988, 1991, Van Valen 1982, Wagner 1986, 1989). The present outline is not meant to argue for the validity of the biological homology concept but to extract its most central assumptions and hypotheses.

The biological homology concept assumes that the body parts, properly defined and delineated, are natural units of biological organization (Wagner 1996).

Hypothesis: organisms consist of quasi-autonomous parts or units

Examples of quasi-autonomous parts or units (QAPs) are cells, vertebrate limbs and insect eyes. In recent developmental literature QAPs have also been called developmental modules. A documentation of the existence of developmental modules can be found in Gilbert et al (1996) and Raff (1996), and a discussion of their evolutionary importance in Bonner (1988) and Wagner & Altenberg (1996). A test that strongly supports quasi-autonomy is the induction of the development of a body part out of its natural context. That this is possible has been shown beautifully for tetrapod limb buds (Hinchliffe & Johnson 1980) and *Drosophila* eyes (Halder et al 1995). These units are able to develop all their defining features in locations of the body where they do not occur in the wild-type or even in organ culture; this demonstrates that the development of these features is locally controlled.

The above hypothesis only implies the existence of developmentally individualized parts of the body but does not say which criteria will be used to define character identity. Analogously, the atomic theory of matter states that matter is composed of corpuscular elements, but does not say which criteria are used to determine particle identity. The determination of such criteria requires a decision about the kind of processes these entities (natural kinds) are assumed to play a role in (Quine 1969). For instance, if the reference processes are chemical reactions, the classification distinguishes between molecules and atoms, and atoms are classified into equivalence classes as chemical elements by their chemical properties. Note that different atoms can belong to the same chemical element (so-called isotopes: ^{12}C and ^{14}C). Similarly, we have to decide which criteria to apply for determining character identity. This can only be done by deciding with respect to which process these entities are expected to act as units

(Wagner 1996). Since the homology concept arose in the context of comparative biology and because the diversity of life is the result of evolutionary change, I propose that homologues are considered as units of phenotypic evolution (Wagner 1995, 1996).

Proposal: homologues are units of phenotypic evolution

According to our current understanding of evolutionary change, phenotypic evolution is based on spontaneous heritable variation. This implies that the properties used to define the identity of units of phenotypic evolution have to be variational properties, i.e. the constraints and the degrees of freedom for natural variation.

Proposal: two quasi-autonomous parts or units are called homologous (i.e. are 'the same' at a certain level of abstraction) if they share certain elements and variational properties (i.e. constraints and degrees of freedom for natural variation)

It has been known for a long time that homologous characters show similar patterns of natural variation. For instance, Schmalhausen (1917; cited after Shubin & Wake 1996) noted that the mesopodial (tarsal and carpal) variation in hynobiid salamanders is strikingly similar to that of other salamanders. This observation has been confirmed many times (Shubin & Wake 1996, Shubin et al 1995, Vogl & Rienesl 1991, Wake 1966, 1991) but the degree of conservation of variational properties is still an open empirical question. Some kinds of mesopodial variants are frequently found in one genus but not in others (Rienesl & Wagner 1992).

This proposal has a number of implications that are testable hypotheses:

Hypothesis 1: the variational properties of characters exhibit historical continuity and coherence. The notions of historical continuity and coherence imply that the defining constraints of a homologue are stable (within limits) during the evolutionary process. It is possible that natural variation may be highly biased, for instance the mesopodial variation in salamanders (Alberch 1983), but this pattern of variation may change erratically from species to species. Such highly unstable variational biases are not useful for defining a unit of evolutionary change since it would not be possible to project their properties to other instances of the homologue.

It is important to note that the stability of the variational properties does not imply immutability. All life is the result of evolution and thus in principle mutable (although sometimes with fatal consequences). Similarly, atoms, the units of chemical transformation, are only stable during chemical reactions but are mutable at high temperatures (e.g. plasma) or may be intrinsically unstable (e.g. radioactive decay). To be useful, one has to expect that the variational properties of homologues are stable under genetic drift, natural selection on the performance of its proper function, and most of the adaptive pressures on other parts of the organism. Loss of function or selection for small body size, however, may have a perturbing effect on the otherwise stable organization of the characters.

Hypothesis 2: quasi-autonomous parts or units form classes of characters that have the same variational properties. This hypothesis has to be invoked to make variational properties meaningful for the definition of character identity. The third hypothesis is an implication that follows from the definition of developmental constraints.

Hypothesis 3: two homologues can only differ in those aspects of their structure which are not subject to their shared developmental constraints. This hypothesis can also be used to test whether a class of characters belong to the same homologue. There has to be a correspondence between the pattern of interspecific variation and the supposed pattern of shared constraints among the characters.

Note that the notion of historical continuity implies that there are character lineages, but it does not logically imply uniqueness, i.e. that all instances of a homologue be part of the same lineage tree. Under this notion parallelisms and atavisms (Hall 1984) can belong to the same biological homologue, even if they clearly do not belong to the same genealogical unit, as required by the historical homology concept.

Hypothesis 4: the variational properties shared by the instances of a homologue are caused by shared developmental mechanisms. In other words, the defining variational properties are developmental constraints (Wagner 1986). It is assumed that the historical continuity of variational properties is caused by inheritance of the developmental mechanisms, and that these developmental mechanisms are maintained by some kind of stabilizing selection. Developmental constraints are inherent to the organismal organization and are not incidental to environmental conditions. This idea is in accordance with the recent theory of natural kinds (Boyd 1989, 1991, Griffiths 1997, Keil 1989), where a natural kind is a unit that exhibits 'causal homeostasis' with respect to the identity-defining properties. The maintenance of the developmental mechanisms provides the causal homeostasis for the variational properties of homologues.

Note that hypothesis 4 does not imply that homologous characters have the same developmental pathway. They do not (de Beer 1971, Hall 1992, Raff 1996, Wagner & Misof 1993). This hypothesis only states that homologues share those developmental mechanisms which cause the shared variational tendencies (constraints and opportunities). Any developmental difference that is irrelevant for the variational tendencies of the character can vary between homologues without any implications on character identity.

Five steps in the identification of biological homologues

Using the above outline I want to propose a research programme to test the biological homology concept.

Identifying two or more putative homologues in a phylogenetically closed group (clade)

An example is the actinopterygian fins and tetrapod limbs. For anatomical characters this step can usually be done on the basis of published anatomical and taxonomic descriptions; however, it is more difficult to determine what the organizational limits of the parts are. Are the entities defined in the anatomical literature developmentally individualized, i.e. are they QAPs? In some cases, like limbs, there are enough experimental results to prove that they are QAPs. In other cases simple perturbation and regeneration experiments may provide clues as to the existence and extent of QAPs. For instance, the fin rays of actinopterygians readily regenerate after partial resection (Goss 1969) with the exception of highly differentiated fin rays (Wagner & Misof 1992). A complete resection of a fin ray, however, does not lead to regeneration[1], which suggests that the fin rays have local form control. Another example of a developmentally individualized character is the hand of mammals, as shown by the observation that double knockouts of *Hoxa11* and *Hoxd11* lead to the loss of the lower arm in mice, but the development of the hand is not prevented (Davis et al 1995).

The putative homologues should also be taxonomically exclusive, i.e. not occur in the same species (e.g. limbs and fins), because taxonomically exclusive characters permit one to contrast different modes of development. Furthermore, by comparing taxonomically exclusive characters it becomes possible to study the evolutionary transformation of one homologue into another, leading into investigations of the mechanisms of evolutionary innovation[2].

Determining the phylogenetic distribution of the putative homologues

This step of course requires a phylogenetic framework of the group that is ideally constructed using an independent set of characters (such as DNA sequences). The objective of this step is to test for phylogenetic continuity and coherence of the putative homologues. It also facilitates the revision of character definitions, if needed. For example, a 'character' that appears and disappears on the phylogeny like a blinking light bulb has no historical continuity and, thus, there is little

[1] A possible exception is caudal fin regeneration in sea needles (Duncker 1905), but this report has not been repeated recently.

[2] Such an investigation is predicated on establishing the character polarity, i.e. which homologue is derived from which.

justification for identifying it as a natural unit of phenotypic evolution. Finally, phylogenetic analysis of the putative characters produces hypotheses about character polarity.

Describing the intra- and interspecific variation pattern of each putative homologue

The purpose of this step is to probe the variational properties of the putative homologues. Interspecific variation positively points to the degrees of freedom of the putative homologue. Invariant features give negative evidence of putative constraints. The study of intraspecific variation further probes for the existence of constraints and degrees of freedom. A feature that is invariant between species but variable within species is clearly not developmentally constrained. A feature that is neither variable among species nor between members of the same species could indicate a developmental constraint (but may still be caused by selection). A definitive picture of the pattern of developmental constraints can only be obtained from a mutation accumulation study, but might be impractical in most cases. A careful analysis of natural variation will nevertheless provide most of the necessary insights. An example is the analysis of mesopodial variation in urodeles by Shubin & Wake (1996).

Describing the development of each putative homologue and determining if modes of development and distribution of homologues are phylogenetically congruent

The goal is to identify developmental differences that might account for the different variational properties of the two homologues. The different modes of skeletogenesis in actinopterygian fins and tetrapod limbs is a good example. Most of the endoskeleton in actinopterygian fins (the radials) develop from a disc of cartilage by secondary individualization (Cubbage & Mabee 1996, Grandel & Schulte-Merker 1998). The existence of a disc is likely an adaptation to the use of the pectoral fin as a stabilizer during larval life. In contrast, the endoskeleton of tetrapod limbs (and sarcopterygian fins?) is derived from a sequence of condensation, segmentation and bifurcation events (Shubin & Alberch 1986). These differences in skeletogenesis likely account for the different variational tendencies of the endoskeleton of actinopterygian fins and tetrapod limbs. Actinopterygian fins always have an anteroposterior series of long bones derived from the anteroposterior extent of the cartilaginous disc, the so-called radials. The variation of the tetrapod limb can be explained from the branching mode of condensations, which form the precursors of the skeletal elements (Müller 1991, Shubin & Alberch 1986).

Once differences in development are identified in exemplary species, like zebrafish and mouse, it is necessary to test whether the taxonomic distribution of modes of development are congruent with the distribution of the putative homologues. For instance, the pectoral fin development of all actinopterygians examined so far—*Salmo salar* (Harrison 1895); *Ammodytes tobianus*, *Gobius minutus*, *Lophius piscatorius*, *Clupea* sp, *Salmo salar* (Derjugin 1910); *Betta splendens* (Mabee & Trendler 1996); *Enchelyurus brunneolus* (Watson 1987); *Chasmodes saburrae* (Peters 1981); *Bathygobius soporator* (Peters 1983); *Anisotromus virginicus* (Potthoff et al 1984)—and even the basal lineages of rayfinned fish, have the larval cartilage disc (*Amia*, *Acipenser* and *Polypterus* cited after Grandel & Schulte-Merker 1998). The aim is to see whether the observed differences exhibit phylogenetic continuity and coherence, qualities necessary for explaining the different variational tendencies of homologues. Furthermore, if the developmental differences are supposed to explain character identity, the taxonomic distribution of the putative character and its underlying developmental cause have to be identical. Differences in this distribution can either show: (a) that the developmental differences are irrelevant to character identity (i.e. variational properties); or (b) that the limits of the character definition are incorrect.

It is certainly impossible to describe the development of each and every species of a clade. With the aid of a good phylogenetic framework, however, it is possible to test whether every instance of a putative homologue is likely to have the developmental feature. This can be done by choosing the most distantly related species that still possess the character. Then one may choose the most 'derived' species with this character and see whether the developmental feature is also found. If these two tests are positive, one can be reasonably confident that all the members that possess the putative homologue share the developmental feature.

Providing and testing a model of how differences in modes of development between putative homologues effect differences in variational tendencies

This step is highly dependent on the biology of the character and cannot be operationalized without detailed knowledge of the biology of the characters. An example is the different mode of hand/foot (autopodium) development in eu-tetrapods (anurans and amniotes) and urodeles. In urodeles there is an absence of a morphologically defined apical ectodermal ridge, variable skeletogenesis which does not follow the metapterygial axis model, and digits which sprout from the limb bud instead of individualizing from a paddle. Finally, the *Hox* gene regulation does not follow the three-phase scheme of limb development (Wagner et al 1998). Of course, whether these differences matter or not is unresolved, since the variation of developmental pathways is not sufficient to reject homology (character identity). The question is, whether the different developmental pathways lead to differences in the developmental constraints. It has been shown experimentally that the difference in autopodial development between urodeles and anurans lead to different tendencies in limb variation that match the pattern of interspecific variation and hence evolution (Alberch & Gale 1985). I conclude that the differences in the developmental programme are associated with a persistent difference in the developmental constraints on natural variation. Thus, the hands of urodeles and eu-tetrapods qualify as different homologues in the sense of the biological homology concept.

The basic idea of this step is to cause experimental perturbations to probe the variational properties of the character. In the case of the Alberch & Gale (1985) experiments, the perturbation was a reduction in the number of cells by a reversible inhibition of mitosis, which was induced to mimic mutational effects. In recent years the number of possibilities thought to interfere with development have increased with the increased knowledge of developmental control genes.

Conclusions

A combination of comparative and experimental methods can be applied to investigate the developmental underpinnings of homologous characters. The guiding idea is that character identification is a hypothesis about shared variational properties in addition to shared history. The critical question is whether homologous characters share developmental mechanisms causing the shared variational properties.

A biological homologue can be considered established if a link can be demonstrated between a developmental difference and different variational tendencies of the two homologues. This link needs to be established at the level of taxonomic congruence as well as at the level of causal efficiency in development. Along the way we need to be prepared to revise the limits of what the QAPs are (what is the part?) and the taxonomic extent of the character definition (Griffiths 1997).

Acknowledgements

I thank Chi-hua Chiu and Kurt Schwenk for reading an earlier version of this manuscript and valuable suggestions. The financial support by National Science Foundation grants IBN 9507466 and IBN 9630567 is gratefully acknowledged.

References

Alberch P 1983 Morphological variation in the neotropical salamander genus *Bolitoglossa.* Evolution 37:906–919

Alberch P, Gale EA 1985 A developmental analysis of evolutionary trend: digital reduction in amphibians. Evolution 39:8–23

Bonner JT 1988 The evolution of complexity by means of natural selection. Princeton University Press, Princeton, NJ
Boyd R 1989 What realism implies and what it does not. Dialectica 43:5–29
Boyd R 1991 Realism, anti-foundationalism, and the enthusiasm of natural kinds. Philosophical Studies 61:127–148
Cubbage CC, Mabee PM 1996 Development of the cranium and paired fins in the zebrafish *Danio rerio* (Ostariophysi, Cybrinidae). J Morph 229:121–160
Davis AP, Witte DP, Hsieh-Li HM, Potter SS, Capecchi MR 1995 Absence of radius and ulna in mice lacking *hoxa-11* and *hoxd-11*. Nature 375:791–795
de Beer GR 1971 Homology: an unsolved problem. Oxford University Press, Oxford (Oxford Biol Readers 11)
Derjugin K 1910 Der Bau und die Entwicklung des Schultergürtels bei den Teleostiern. Z Wiss Zool 96:572–653
Donoghue MJ 1992 Homology. In: Keller EF, Lloyd EA (eds) Keywords in evolutionary biology. Harvard University Press, Cambridge, MA, p 171–179
Duncker G 1905 Über Regeneration des Schwanzendes bei Syngnathiden. Roux's Arch Entw Mech 20:30–37
Gilbert SF, Opitz JM, Raff RA 1996 Resynthesizing evolutionary and developmental biology. Dev Biol 173:357–372
Goss RJ 1969 Principles of regeneration. Academic Press, New York
Grandel H, Schulte-Merker S 1998 Paired fin development in the Zebrafish (*Danio rerio*). Int J Dev Biol, in press
Griffiths PE 1997 What emotions really are. University of Chicago Press, Chicago, IL
Halder G, Callaerts P, Gehring WJ 1995 Induction of ectopic eyes by targeted expression of the *eyeless* gene in *Drosophila*. Science 267:1788–1792
Hall BK 1984 Developmental processes underlying heterochrony as an evolutionary mechanism. Can J Zool 62:1–7
Hall BK 1992 Evolutionary developmental biology. Chapman & Hall, London
Harrison RG 1895 Die Entwicklung der unpaaren und paarigen Flossen der Teleostier. Arch Mikr Anatomie 46:500–578
Hinchliffe JR, Johnson DR 1980 The development of the vertebrate limb. Oxford University Press, New York
Keil FC 1989 Concepts, kinds and cognitive development. MIT Press, Cambridge, MA
Mabee PM, Trendler TA 1996 Development of the cranium and paired fins in *Betta Splendens* (Teleostei: Percomorpha): intraspecific variation and interspecific comparisons. J Morph 227:249–287
Müller GB 1991 Evolutionary transformation of limb pattern: heterochrony and secondary fusion. In: Hinchliffe JR, Hurle JM, Summerbell D (eds) Developmental patterning of the vertebrate limb. Plenum Press, New York, p 395–405
Osche G 1973 Das Homologisieren als eine grundlegende Methode der Phylogenetik. Aufsätze Reden Senckenberg Naturf Ges 24:155–166
Peters KM 1981 Reproductive biology and developmental osteology of the florida blenny, *Chasmodes saburrae* (Perciformes: Blenniidae). Northeast Gulf Sci 4:79–98
Peters KM 1983 Larval and early juvenile development of the frillfin goby, *Bathygobius soporator* (Perciformes: Gobiidae). Northeast Gulf Sci 6:137–153
Potthoff T, Kelley S, Moe M, Young F 1984 Description of porkfish larvae (*Anisotremus virginicus*, Haemulidae) and their osteological development. Bull Mar Sci 34:21–59
Quine WV 1969 Natural kinds. In: Quine WV (ed) Ontological relativity and other essays. Columbia University Press, New York, p 114–138
Raff R 1996 The shape of life: genes, development and the evolution of animal form. Chicago University Press, Chicago, IL

Rienesl J, Wagner GP 1992 Constancy and change of basipodial variation patterns: a comparative study of crested and marbled newts — *Triturus cristatus, Triturus marmoratus* — and their natural hybrids. J Evol Biol 5:307–324
Roth VL 1984 On homology. Biol J Linn Soc 22:13–29
Roth VL 1988 The biological basis of homology. In: Humphries CJ (ed) Ontogeny and systematics. Columbia University Press, New York, p 1–26
Roth VL 1991 Homology and hierarchies: problems solved and unresolved. J Evol Biol 4:167–194
Shubin N, Alberch P 1986 A morphogenetic approach to the origin and basic organization of the tetrapod limb. Evol Biol 20:319–387
Shubin N, Wake D 1996 Phylogeny, variation and morphological integration. Am Zool 36:51–60
Shubin N, Wake DB, Crawford AJ 1995 Morphological variation in the limbs of *Taricha granulosa* (Caudata: Salamandridae): evolutionary and phylogenetic implications. Evolution 49:874–884
Van Valen L 1982 Homology and causes. J Morphol 173:305–312
Vogl C, Rienesl J 1991 Testing for developmental constraints: carpal fusions in urodeles. Evolution 45:1516–1519
Wagner GP 1986 The systems approach: an interface between development and population genetic aspects of evolution. In: Raup DM, Joblonski D (eds) Patterns and processes in the history of life. Springer-Verlag, Berlin, p 149–165
Wagner GP 1989 The biological homology concept. Ann Rev Ecol Syst 20:51–69
Wagner GP 1995 The biological role of homologues: a building block hypothesis. Neues Jahrb Geol Palaeontol Abh 195:279–288
Wagner GP 1996 Homologs, natural kinds and the evolution of modularity. Am Zool 36:36–43
Wagner GP, Altenberg L 1996 Complex adaptations and the evolution of evolvability. Evolution 50:967–976
Wagner GP, Misof BY 1992 Evolutionary modification of regenerative capability in vertebrates: a comparative study on teleost pectoral fin regeneration. J Exp Zool 261:62–78
Wagner GP, Misof BY 1993 How can a character be developmentally constrained despite variation in developmental pathways? J Evol Biol 6:449–455
Wagner GP, Khan PA, Blanco MJ, Misof BY, Liversage RA 1998 Evolution of *Hoxa-11* expression in amphibians: is the urodele autopodium an innovation? Am Zool, in press
Wake DB 1966 Comparative osteology and evolution of the lungless salamanders, family Plethodontidae. Mem South Calif Acad Sci 4:1–111
Wake DB 1991 Homoplasy: the result of natural selection, or evidence of design limitations? Am Nat 138:543–567
Watson W 1987 Larval development of the endemic Hawaiian blenniid, *Enchelyurus brunneolus* (Pisces, Blenniidae, Omobranchini). Bull Mar Sci 41:856–888

DISCUSSION

Meyer: This may only be a semantic issue, but would you say that digits 3 and 4 in salamanders are convergent?

Wagner: Homology is a hierarchical concept, and I was talking about the level of the autopodium. There is genetic evidence that the autopodium is a quasi-autonomous unit: in *Hoxa11* and *Hoxd11* knockout mice the lower arm is absent but there is a more or less fully developed hand. Therefore, the autopodium is a self-contained developmental unit that can develop without the presence of the proximal elements. This, in my view, is an appropriate level of analysis. I'm not sure to what degree the different digits are individualized, and this is necessary to know to be able to homologize them one by one. I would expect the programme that leads to the formation of phalanges to be the same programme that was used in the ancestral eu-tetrapods, because the hypothetical ancestor of urodeles still maintained two fingers. In this respect they are homologous. My scenario only explains why there is a different mode of digit formation compared with other tetrapods. If they are innovations it also explains why the urodele autopodium development is more plastic (because new forms tend to be less canalized than older characters). We only find this variation in hand development in urodeles, and so it is consistent with the notion that they are a new structure.

Greene: Have you looked at amphiumas, among which there is one species with one toe, one species with two toes and one species with three toes?

Wagner: Shubin & Alberch (1986) have looked at the species with two toes. Contradictory to the model they pushed in their paper, they drew a connection from the fibula to the basal commune in their schematic diagrams. This is consistent with the notion that the last remaining toe of amphiuma is derived from the metapterygial axis, i.e. homologous to digit 4 in eu-tetrapods.

Carroll: Would you forecast that we should be able to find a series of fossils which show that the ancestral urodeles only had two digits?

Wagner: That would be the prediction, although I wouldn't be so bold to say that we will find them. I expect that they will be small.

Carroll: But adult hynobiids always have normal digits, as do cryptobranchids. Only amphiumids, proteids and some plethodontids have fewer digits.

Wagner: I'm not claiming that amphiumids are the ancestors of the other urodeles because their phylogenetic positions are incompatible with that hypothesis. All I'm saying is that they have the propensity in certain ecological situations to undergo a reduction in limb size, as do lepospondyls, in which limb reduction is often accompanied with body elongation.

Rudolf Raff: You mentioned the term 'developmental constraint', but in the examples you quoted, you seemed to use it to mean developmental mode. Could you clarify the term for me, because I don't see why something should necessarily be under developmental constraint if it has just continued for 200 million years without change?

Wagner: No, that's not what I mean. By the term 'developmental constraint' I mean any bias or limitation on the production of phenotypic variation caused by the developmental mode. If I wanted to prove that two homologues are two different biological entities that have different variational tendencies, I would have to demonstrate that these differences in variational tendencies are caused by differences in developmental mode. However, developmental variation can be irrelevant for variational tendencies, and the question would then be, what developmental differences are relevant? These would only be those differences in developmental mode that are stable and affect the limitations and degrees of freedom of natural phenotypic variation. For example, in the cartilaginous disc of actinopterygians the mode of skeletal development is to just chop it up into a series of elements. This may explain the tendency to create a series of skeletal elements in an anteroposterior dimension, rather than proximodistal (as is typical in tetrapod limbs). Therefore, I'm not focusing on the stability of the disc, but rather its effect on the kinds of osteological patterns that can be created.

Rudolf Raff: Would it be more fruitful to speak of different developmental mechanisms rather than different developmental constraints?

Wagner: But different developmental mechanisms are only important for character identity if they have consequences for the constraints and opportunities of phenotypic variation. For example, in the process of digit reduction certain variants are more likely in frogs versus salamanders, and this is a consequence of developmental constraints. It is not that the cellular mechanism is different, but rather there is a difference in the developmental sequence that affects character variation.

Wray: I have a general question about the biological homology concept. This is clearly a concept in a research programme that is geared towards understanding homology in morphological structures, but there are homologous genes, homologous cell types and homologous developmental pathways. Is there such a thing as homology for those entities? Because they're not compatible with your research programme.

Wagner: I'm interested in morphological characters and vertebrate development, and I'm trying to make sense of homology at the level of anatomical structures in vertebrates. I'm cautious not to extend these concepts to other areas. Similar ideas may be important for gene networks, for example, but I wouldn't feel confident in saying that they are.

Wray: What about extending them to homologous genes?

Wagner: The conceptual problem is less stringent for genes because we know the mechanistic basis for the continuity of information. Due to the semiconservative mode of DNA replication there is even a material continuity from one generation to the next. It is more difficult for morphological structures and behaviours because there is no continuity of the structures themselves from generation to generation. I suggest that there is an invariant developmental mechanism that is responsible for the continuity of character identity, even if the characters themselves have no material continuity. Another reason why I'm hesitant to overextend this notion is that if you want to talk about mechanisms, you are bound to the biology of the character you are studying, and another type of character may have a different biology.

Meyer: I would like to argue with you about the importance of character status as opposed to character identity. Your interpretation is that character identity is just the frosting on the cake and is not fundamentally important, whereas I would argue that these new homologies you described arise as autapomorphies that are the differences which distinguish species.

Wagner: My point is that in order to make statements about the genealogical relationships within a particular group, we must apply the cladistic terms to character states as well as the presence or absence of characters. This is fine as a tool, but it would be futile to try to develop a mechanistic understanding of character identity that also encompasses differences in character state.

Meyer: It's just another level, because one can argue that the wings of a bat and a bird are not homologous, but that the forearms of a bat and a bird are. This is a difference in character state when one looks at different levels of phylogenetic inclusiveness.

Wagner: One way to get out of this circular argument is to say that if the transformations that led to the evolution of a bird wing and a bat wing are associated with the acquisition of new constraints, then a new level of homology is added. In that sense there would be different hierarchical levels of homology because there are different levels of constraints.

Meyer: But then you would need to argue that by looking at character identities at this deep level you would have a higher chance of discovering these developmental constraints, because presumably they are more fundamental and more different than at a less inclusive level of phylogenetic comparison.

Wagner: But the goal is not to do phylogenetic analysis, but rather to understand how character identity is changing and how innovations arise. I look at character homology in the same way that people look at the species concept. Speciation needs to be mechanistically explained. Therefore, I call this the biological homology concept to relate it to the biological species concept, which also moved away from taxonomic problems towards the understanding of population genetic mechanisms of the origin of species. We need to get away from taxonomic problems and try to focus our attention on those questions that are more relevant for understanding the origin of new characters.

Meyer: But your approach is not the same as the biological species concept approach because you are not using information at the population level, rather your approach is more typological.

Wagner: It is not typological because I am analysing developmental mechanisms that are responsible for the observed variational properties of characters.

Striedter: You mentioned that changes in the mode of development are important in terms of homology only if they create different constraints, and

Rudolf Raff in his presentation gave examples of where homologous structures develop differently through different modes. Would you therefore speculate that in all those cases the new modes of development are subjected to these constraints? I find this hard to believe.

Wagner: I don't believe that either. It is clear that developmental pathways can vary without affecting the adult character. This is biologically plausible in sea urchins, for example, because they undergo metamorphosis, and metamorphosis is a way of decoupling larval/embryological morphology from adult morphology. Therefore, it is easy to radically change early development because the metamorphosis mechanisms are conserved.

Rudolf Raff: The changes are deeply ingrained, and the convergence pattern of the developmental hourglass takes place before metamorphosis, so you can't say that metamorphosis is a decoupling event.

Wray: Regeneration and asexual reproduction are other counter examples in which non-embryonic processes determine the field sizes, the number of cells, etc.

Wagner: But I would argue that in these examples a particular anatomical end product is not due to the developmental pathway, but rather the self-stabilizing property of the end product. If this is the case, and one has to assume that there are self-stabilizing mechanisms if characters are made out of living cells and have cell turnover and growth, since the anatomical configuration doesn't change there must be something actively maintaining this anatomical configuration in spite of growth and cell turnover.

Wray: What is the null hypothesis for self-stabilizing activity? I would like to know what you are contrasting it with, and how you would test it.

Wagner: When self-stabilizing activity is present perturbations in the developmental pathway don't percolate into the end product. The end product can only be altered if there are changes in the tissue interactions that are necessary for the self-maintenance of the end product.

Wilkins: By the time the end product is formed the process is complete.

Wagner: No, it is not complete because there is still cell turnover.

Wilkins: If there are many different ways of reaching the same end product, surely one can't say that the end product is exerting some sort of control over itself.

Wagner: But pathology is the breakdown of self-maintenance of the organs.

Wilkins: I would say that's a different issue.

Wagner: It's not a different issue because it is necessary to explain why the end product is a morphogenetically stable entity.

Nicholas Holland: You seem to present this research programme as a universal calculus for the study of body part homologies by looking at shared developmental constraints and mapping modes of development onto independently derived cladograms. What you're doing works well for vertebrates and invertebrate deuterostomes, but there is a limited fossil record for many invertebrates, and for the protostomes there is as yet no reliable, independently derived cladogram at the phylum levels on which to map the modes of development. Therefore, those who work on protostome invertebrates are going to have to use something else, at least for now.

Wagner: All I am saying is that if you want to do this research programme, you have to have a good phylogeny. If you don't have it, you should try something else.

Nicholas Holland: But you're almost telling us that we can't look at most of the animal kingdom.

Wagner: You can, but it depends on what your goals are.

Nicholas Holland: My goal is to reconstruct the history of life. I would sell my soul to the devil to understand this.

Greene: Don't do that yet! People are struggling to understand this, but progress is being made.

Nicholas Holland: It's not being made with this particular methodology.

Wagner: But my goal is not to reconstruct phylogeny. A phylogenetic framework is required to perform these analyses.

Hinchliffe: Your hypothesis is intriguing, but I'm not entirely convinced by the evidence. In particular, according to the theory, the prospective autopod (from which digits form) is present in the urodeles (even if repositioned) as it is in anurans. So one might expect *Hoxa11* expression to be absent from the future autopod in both *Xenopus* and the urodele, *Notopthalmus*. Yet dramatic differences are illustrated between *Hoxa11* expression in *Xenopus* and *Notopthalmus* prospective autopods. Surely, according to your hypothesis shouldn't one expect this expression to be reasonably similar?

Wagner: The differences in gene expression between urodeles and eu-tetrapods refer to two events. (1) The initial stages of digit formation of digits 3, 4 and 5, which involves a sort of 'digit bud', is quite different from digit formation in eu-tetrapods. I take this as evidence that the development of these posterior digits is different from that in eu-tetrapods, and thus may indicate that these digits are 'neo-morphs'. (2) *Hoxa11* is expressed during the elongation of all digits. This has not been described for eu-tetrapods, but gene expression studies have always ignored the stages of development in which phalanges are added to the embryonic digits. We have preliminary evidence that this expression pattern is also present in toads, but it is not known whether it also occurs in mammals and birds. Hence, there are few comparative data about *Hox* gene expression variation in the limb, and it is difficult to say what one would expect to observe.

Akam: I love your research programme, but can't you call it something other than homology, because that's just making it more difficult to understand?

Wagner: It could be called a research programme for the origin of character identity.

Reference

Shubin N, Alberch P 1986 A morphogenetic approach to the origin and basic organization of the tetrapod limb. Evol Biol 20:319–387

Homology and homoplasy: the retention of genetic programmes

Axel Meyer

Department of Biology, University of Konstanz, 78457 Konstanz, Germany

Abstract. Homology describes the inevitable evolutionary phenomenon that the similarity of structures among different organisms is due to the commonality of their descent. This continuity of information is maintained in evolutionary lineages in terms of genes and developmental mechanisms and will retain 'sameness' and retard, funnel and direct evolutionary diversification. Analogous 'sameness' is said to be due to independent, convergent evolution, and also involves similarity of function; the latter is not a necessary condition for structures to be identified as homologous. Here, I suggest that the biological basis for these seemingly disparate kinds of 'sameness' in evolution may in some, or even most, instances not be all that different and may be based on the same principle—the long evolutionary retention of genes, gene interactions and developmental mechanisms. Evolution might recycle and re-recruit similar mechanisms repeatedly during its course, and it often makes do with what is already available to it rather than to newly evolve or reinvent many gene interactions and developmental mechanisms repeatedly. Apparently there is no, or only a negligible, 'genomic cost' or even a selective advantage to maintaining genes and developmental mechanisms for long evolutionary periods of time, even if they are not continuously used in all members along an evolutionary line. Therefore, the biological basis of both homologous traits (those that are evolutionarily always expressed) and homoplasious traits (those that are not always 'on', but are 're-awakened' during evolution) might not be so different, and the distinction between homology and some forms of homoplasy may be somewhat artificial.

1999 Homology. Wiley, Chichester (Novartis Foundation Symposium 222) p 141–157

How different are homology and homoplasy?

Why should one be interested in homology? Because it is one principle, maybe *the* unifying principle, of evolution. By understanding homology, it might be presumed, we will better understand some of the rules by which evolution proceeds, about regularity of processes in evolution, about understanding patterns and trends in evolution and, concomitantly, about diversification of form. The fact that homologous structures exist provides one of the strongest lines of evidence for evolution, but so does convergent evolution in the form of analogy, parallel evolution, reversals and atavisms (reviewed in Futuyma 1998).

Homoplasious phenotypes also attest to the strength of selection, or alternatively internal constraints, to sculpt similar phenotypes in response to similar selection pressures and to re-express ancient, retained, developmental programmes. Therefore, to improve our understanding of the predominant mechanisms by which evolution proceeds, and how biological diversity is achieved and maintained, we need to investigate the biological basis of both homology, and the various forms of homoplasy, such as convergent evolution (analogy), parallel evolution, reversals and atavisms.

Homology, as is well known, is a pre-Darwinian and pre-phylogenetic concept that is already more than 150 years old (Owen 1848, reviewed in Panchen 1994). The theory of homology continues to be debated and the concept itself has evolved over time. Today, it usually implicitly or explicitly involves the recognition of similarity of structure in organisms due to shared recent common ancestry. The recognition and definition of homology are clearly dependent on phylogenetic knowledge. But phylogenetic continuity is only a necessary, but not sufficient, criterion of homology. None the less, sometimes homology is even narrowly defined in a cladistic sense as synapomorphy—as the persistence of traits in their various transformed states (Nelson 1994). However, this is clearly an overly restrictive and somewhat circular definition because the status of a character state as synapomorphy (i.e. homology) or symplesiomorphy will depend on the set of taxa included in a phylogenetic analysis (Wake 1999, this volume).

Typically, there are four criteria by which one can recognize homology: (1) similarity of structure; (2) position (anatomical relationship); (3) phylogenetic continuity; and sometimes a fourth criterion is invoked—sameness of the underlying developmental basis of two similar structures. Note that function is not, and never has been, a defining characteristic of homology. All of these criteria have to be met, if a single one is not, the structures should not be considered homologous. Homology remains a contentious and difficult concept (Hall 1994, Wake 1994), and one might argue that it is not useful to continue to argue only about definitions, but that the underlying basis of 'sameness' in nature is what should be studied (Wake 1999, this volume).

If similar structures are not considered homologous then they are often considered to be one of several forms of homoplasy (convergence, parallelism or reversal). Lankaster (1870) coined the term 'homoplasy' as the appearance of sameness resulting from independent evolution, defined as derived similarity that is not synapomorphic. Homoplasy is usually divided into three more or less arbitrary classes.

(1) Convergent evolution (or analogy): recognized through superficially similar features that evolved independently and arose ontogenetically by different pathways.

(2) Parallel evolution: similar developmental modifications that reappeared independently, but were not present continuously in all members of an evolutionary lineage. Parallel evolution occurs among closely related organisms, due to parallel evolution in structures likely to be formed by identical or similar developmental mechanisms.
(3) Reversals, atavisms and rudiments: a 'return' from an advanced character state to a more 'primitive' or ancestral state. It is not clear whether atavistic structures are formed by the same or similar developmental mechanisms as the original structures were that they resemble.

Homoplasy, it might be contended, is even more common in evolution than homology. One piece of evidence in favour of this tenet is that 'phylogenetic noise' is typically more prevalent than phylogenetic information in taxonomic data sets. If homoplasy is extremely common and convergence is common, then one might suggest that some of this commonality should be able to teach us something about regularity, rules and possibly processes by which these patterns of homoplasy are brought about in evolution (Sanderson & Hufford 1996). Similarly, if one sees convergence as the flipside of homology, then, almost as a by-product, one is also going to learn about homology by studying convergence and parallelism.

Therefore, it might be as important to understand non-homology as homology. Both might reveal which kinds of internal and external forces constraint shape and possibly even direct the diversity or commonality ('sameness') of shapes in evolution.

'Functional' and 'partial' homology

The debate about homology resurfaced anew since the recent publication of comparative developmental data about the expression patterns of apparently homologous genes in seemingly non-homologous structures. *Pax-6* expression in precursor cells of various kinds of light receptors in different phyla is one of the most striking examples of these kinds of results (e.g. Halder et al 1995). This gene is switched on in many light-detecting morphological structures that, based on evolutionary, structural and developmental criteria would by most biologists clearly not be considered to be homologous. These data document a surprising degree of conservation in evolution and might even necessitate a re-evaluation of the concept of homology (Abouheif et al 1997).

The compound eyes of insects and the camera eyes of vertebrates surely are not homologous structures based on the criterion of evolutionary continuity since eyes such as these evolved independently many times over in the history of animals. None the less, these eyes were categorized by some as homologues on the basis of

Pax-6 expression data. The findings show that apparently homologous genes can be expressed, possibly even in homologous networks of genes (Abouheif 1999, this volume), in phenotypically seemingly non-homologous structures. The recognition of genes as homologous requires knowledge of the phylogenetic relationships of all members of a gene family (e.g. Zardoya et al 1996). The expression of *Pax* genes in phenotypically differing kinds of light receptors raised the issues of 'levels of homology', 'partial homology' and 'functional homology' (Abouheif et al 1997).

As mentioned before, similarity of function has never been part of the definition or one of the criteria of homology, but in the developmental literature similarity of function is often erroneously used to falsely identify genes and structures as 'homologous' based solely on their similarity of expression pattern. This similarity is interpreted to imply that the structures in which these 'functionally homologous' genes are expressed are also always evolutionarily homologues. They may or may not be homologous, but expression patterns alone are insufficient evidence for homology of structures.

'Partial homology' describes a situation where genes are homologous, but the structures in which they are expressed are not. A probable example of this phenomenon is the expression of *Pax* genes and different light receptors in diverse animal phyla. It should be clear from the *Pax-6* example that homologous genes can 'make' non-homologous structures. Once a contentious concept, partial homology did, as a result of these kinds of new comparative developmental data, become relatively widely excepted (Wake 1999, this volume).

These new findings from evolutionary developmental biology raise the converse question: can partial homology exist where the structures are homologous, but the genes that are expressed in their precursors are not? The answer is probably yes; non-homologous genes, gene networks and developmental mechanisms can make structures that are typically considered to be homologues. If homology of genes were a necessary criterion of homology (the fourth criterion, see above) then all homologous structures made by non-homologous genes could not be considered to be homologues. This is clearly not the case in the opinion of many researchers.

The biological basis of 'sameness'

Biological explanations must be sought for phenomena such as stasis, modularity, preservation of design, latent homology and directionality of evolution (Wake 1999, this volume). The exciting new data from comparative developmental biology document an unexpected degree of conservation of genes and genetic programmes. Homologous genes are retained for extensively long evolutionary time spans, and they can be expressed during the development of structures that, based on phenotypic criteria, would not be considered to be homologous but analogous. These kinds of data might shed light on the underlying basis of biological 'sameness', both in the form of homology or various forms of non-homology. If both homology and homoplasy result in 'sameness' (Wake 1999, this volume) between organisms, maybe the biological bases of both kinds of sameness might not be so different after all? One might ask, what are the biological bases for similarity of features whether they evolve independently or not? The ubiquity of homoplasy in the form of parallel and convergent evolution in nature poses the question of what can be learned from knowledge of the developmental systems about patterns in evolution and about the origin of novel features.

In this context it should be remembered that all species are mixtures (mosaics) of ancestral and derived states of characters. Mosaic evolution refers to the fact that evolution proceeds at different rates for different characters within one organisms and in evolutionary lineages. Species do not evolve as a whole but by piecemeal: many of their features evolve quasi-independently. Incidentally, this implies that because of mosaic evolution we cannot call species primitive or advanced and only particular characters are basal or derived. The fact that evolution proceeds this way can best be seen among closely related species that only differ in some but not all traits.

The retention and 're-awakening' of developmental mechanisms in convergent evolution

With more than 2000 species, fish in the cichlid family are one of the most successful groups of vertebrates. Cichlids were recently recognized to be a major example of extensive and repeated parallel evolution. In East Africa, the centre of their distribution, they form adaptive radiations with several hundred endemic species in each of the three East African great lakes: Lakes Tanganyika, Malawi and Victoria. Conspicuous parallels are found in terms of various phenotypic traits such as morphology, striking ecological specialization, colour patterns and behaviour in these three major species flocks. From molecular phylogenetic work it is now known that the relatively young and genetically rather homogeneous Lake Malawi and Lake Victoria cichlid flocks are monophyletic (or oligophyletic in the case of Lake Malawi, Meyer et al 1990). Their single ancestral lines are derived from one of the 11 lineages of cichlids that form the much older and genetically more diverse species flock from Lake Tanganyika (reviewed in Meyer 1993; Fig. 1).

Before the advent of molecular phylogenetic data, a different phylogeny of cichlids from these three lakes had been proposed that was based on phenotypic traits, namely that the often astonishing similarity of morphologies between cichlid species occurring in each of the three lakes (Fig. 2) are indicative of close evolutionary relationships. Hence, it was assumed that each of the three radiations of cichlids from East Africa had multiple ancestors that lived in the other lakes.

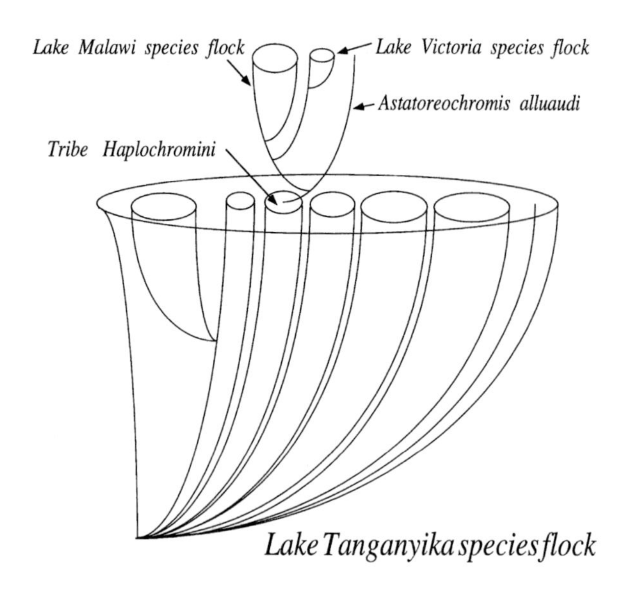

FIG. 1. Cartoon of the phylogenetic relationships of cichlid fish among the species flocks of the three great East African lakes. The species flock in Lake Tanganyika is composed of several, mostly endemic, major lineages, one of which is the tribe Haplochromini. The ancestors of the Lake Malawi and Lake Victoria adaptive radiations are derived from this lineage. The degree of genetic diversity (horizontal axis) and age (vertical axis) of the three species flocks differ significantly, indicated by the differences in the size of the vesicles that symbolize the three species flocks.

Similar morphologies and lifestyles such as mollusc crushing or algae scraping — e.g. from species such as *Tropheus* from Lake Tanganyika and *Pseudotropheus* from Lake Malawi (Fig. 2) — were seen as indications of close evolutionary relationships. The molecular phylogenetic data show that this is not the case, but rather that generalized ancestral species for each of the flocks (with the possible exception of the Lake Tanganyika radiation, which was founded by several ancestral lineages) gave rise repeatedly and independently to similar phenotypes in parallel in the three lakes either in response to similar selection pressures and/or due to retained ancestral developmental constraints or the inherit increased 'evolvability' (Kirschner & Gerhart 1998) of cichlids. Likewise, the morphologically highly specialized species from each of the younger flocks are more closely related to each other than to often phenotypically similar species in the other lakes.

The oldest of the East African cichlid lineages are probably not more than 10 million years old, yet they have attained a marvellous diversity of shapes, sizes and colours. Even more impressive is the fact that the Lake Victoria species flock of cichlids, with about 500 endemic species, might be as young as 14 000 years, yet it has the same array of ecological and behavioural diversity as the much older Lake

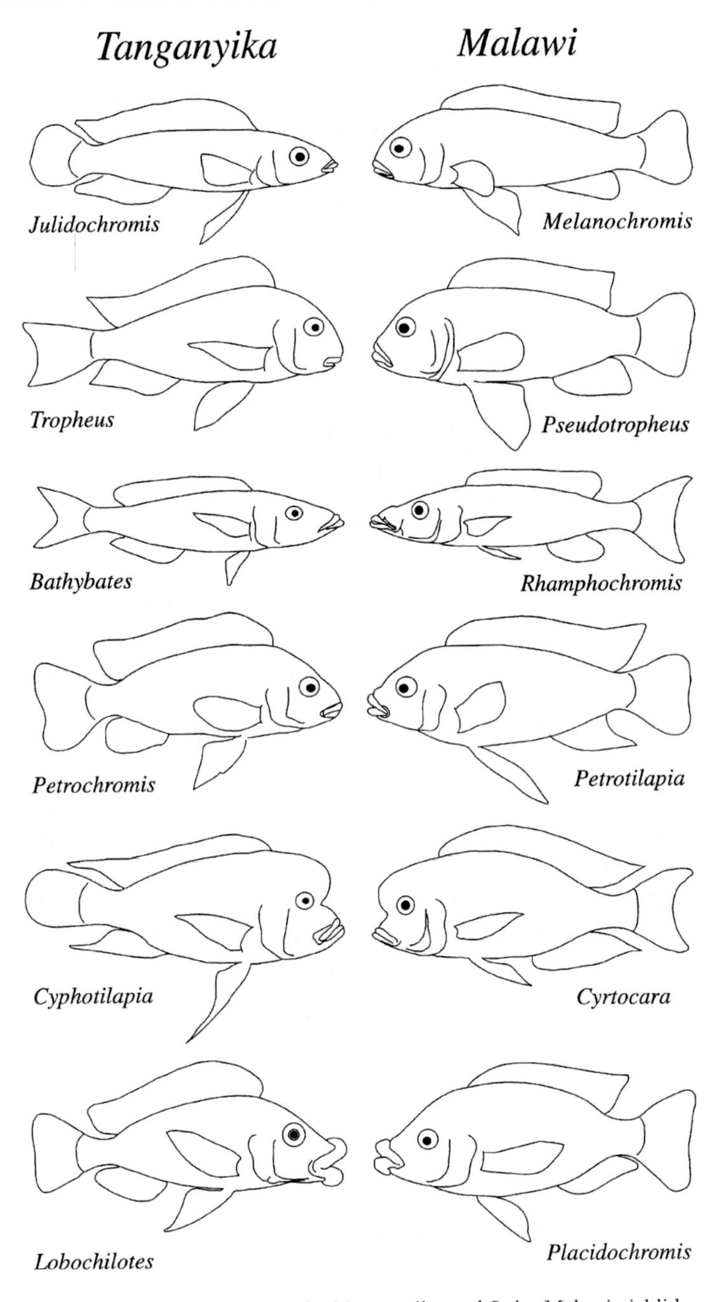

FIG. 2. Juxtaposition of endemic Lake Tanganyika and Lake Malawi cichlids to show the phenotypic similarity of the species that also perform ecologically equivalent roles in their respective species flocks.

Tanganyika cichlid flock. There are algae-scraping, fish-eating, scale-scraping, mollusc-crushing, zooplankton-eating, etc. specialists in each of the three species flocks. Among these relatively closely related species we hence observe ultra-fast speciation rates, with concomitant rapid phenotypic diversification. Yet, despite the fact that rapid speciation in cichlids appears to be the norm, the molecular data also demonstrate cases in some lineages of relatively long periods of stasis where morphological diversification appeared to have come to a relative standstill (Sturmbauer & Meyer 1992).

In terms of understanding how patterns in evolution came about it might be informative to study whether the developmental mechanisms that create almost identical phenotypes in this prominent case of parallel evolution are the same or whether different developmental mechanisms led convergently to the same morphological end in each of the three adaptive radiations. Such comparative approaches among closely related species with divergent morphologies, or among distantly related species with parallel morphologies, require phylogenetic knowledge. It might be surmised that this astonishing similarity of phenotypic traits in cichlids (Fig. 2) evolved in parallel possibly by using the same, retained developmental mechanisms, rather than by the repeated invention of developmental mechanisms.

Parallel evolution and the retention of genes in the context of sexual selection

Another example of gene retention and parallel evolution involves a sexually selected trait in fish, i.e. the 'sword' of males in some fish in the genus *Xiphophorus* (Fig. 3). Sexual selection favours coloured, ventrally elongated caudal fins, or swords in males of the fish in this genus. Males with longer swords are preferred by females over males of equal size with shorter swords (Basolo 1990). Even females of some species in which the males do not have swords seem to prefer males with artificial swords and heterospecific males with swords. Females of the basal sword-less species *Xiphophorus maculatus* prefer males with swords although their conspecific males do not posses them (Basolo 1990; Fig. 3). Viewed in the context of the traditional phenotypically based phylogeny (Fig. 3) this led to the suggestion that the evolution of swords evolved in response to an evolutionary earlier bias in females to mate with males that posses this trait, since the preference in females existed before the trait itself (Basolo 1990; Fig. 3).

For several reasons this 'pre-existing bias' hypothesis may not hold completely because, for example, the males of *Xiphophorus xiphidium* already possess a ventral, albeit colourless, extension of the ventral caudal fin rays. Moreover, the reconstruction of the evolution of the sword within the genus based on a molecular phylogeny of fish of this genus (Meyer et al 1994; Fig. 4) suggests that the sword evolved early in the evolution of the genus (black bar at the base of the

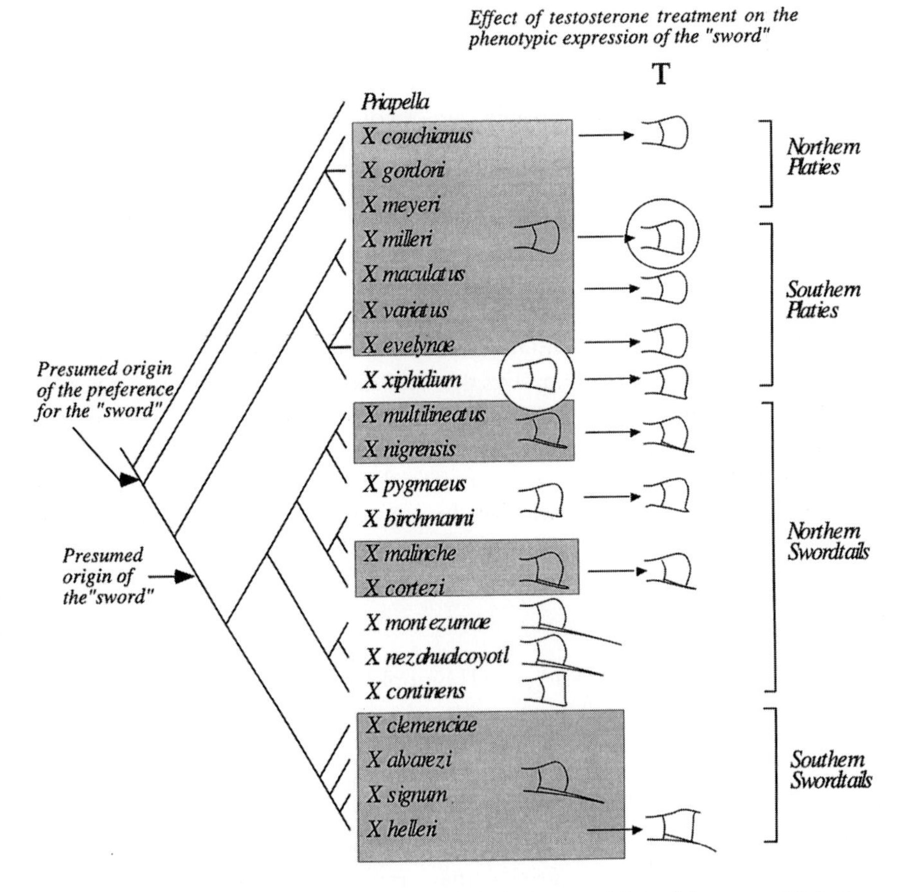

FIG. 3. Traditional (based on phenotypic traits) phylogeny of fish of the genus *Xiphophorus*. The typical size of the ventral caudal fin extension in males of each species is indicated in natural populations and after treatment with testosterone.

genus in Fig. 4), was then lost at least once (open bar in Fig. 4) and re-evolved at least twice in *Xiphophorus clemenciae* and *X. xiphidium* (Fig. 4). Furthermore, it is important to note that there are several other species of poeciliid fishes (e.g. in *Poecilia petenensis* and *Poecilia reticulata*) and many other families of fish where males have notable elongations of the most ventral rays of their caudal fins. Often, these ventral elongations can, through artificial selection, lead to extremely pronounced swords in some, naturally typically sword-less, species (e.g. *P. reticulata*). It should also be mentioned that in several species of *Xiphophorus* there is, sometimes extensive, variation in natural populations in terms of the phenotypic expression of swords (Meyer 1997).

It would appear that 'sword' genes might be ancient in poeciliids and other fish, and are variously expressed, i.e. repeatedly lost and regained during evolution. Experiments that involve the treatment of males in several species of *Xiphophorus* with testosterone show that in some cases swords could be induced in species whose ancestors apparently never had them (Fig. 3), based on the interpretation of the traditional hypothesis (Basolo 1990, 1991). The molecular phylogeny for the genus (Fig. 4) suggests that swords were repeatedly lost and regained within the genus. These observations might suggest that, since in naturally sword-less species (e.g. *Xiphophorus milleri* and *X. maculatus*, Figs 3 & 4) swords can be induced through testosterone treatment, the genetic and developmental machinery to produce swords might have been retained and 're-appeared' when selected for by sexual selection (Fig. 4). The molecular phylogeny makes the interpretation of these testosterone experiments much more consistent. Based on the traditional phylogeny it would be difficult to understand why sword-like extensions of the ventral rays in the caudal fin could be induced through testosterone if ancestral species did not have swords and they did not retain the 'sword gene', as is suggested by the molecular phylogeny.

Conclusions

'Sameness' in the form of homology and homoplasy may not be all that different, or may be at least partly caused by the same mechanism: the ubiquitous evolutionary retention of genetic potentiality. The retention of genes, genetic networks and embryological pathways may be a prevalent mechanism by which stasis in morphology is attained, and it may also be one mechanism by which similarity in the form of homologous and homoplasious structures is achieved. Instead of inventing similar structures entirely anew from the genes upwards through transcription regulation, gene networks and morphogenetic mechanisms, evolution often appears to draw on 'old' genes and mechanisms that are 'recycled' and re-used for different purposes throughout evolution (Gerhart & Kirschner 1997). The idea that evolution may be haphazard and often makes do with what it has at its disposal is essentially a Jacobian idea (Jacob 1977). Evolution is a tinkerer and it will work with what it has at its disposal — it might recycle 'old' genes and genetic and embryological pathways when they are required in the same ontogenetic or ecological context to solve an ecological problem.

If developmental mechanisms (brought about by conserved developmental genes and their interactions) are retained for long evolutionary times spans (on the scale of the age of phyla in the case of *Pax-6* genes) it may be at a low or no 'genomic cost'. The evolutionary forces that maintain genes and genetic interactions will presumably have to outweigh those mechanisms (such as mutation, selection, etc.) that will lead to the deterioration of genes once they are

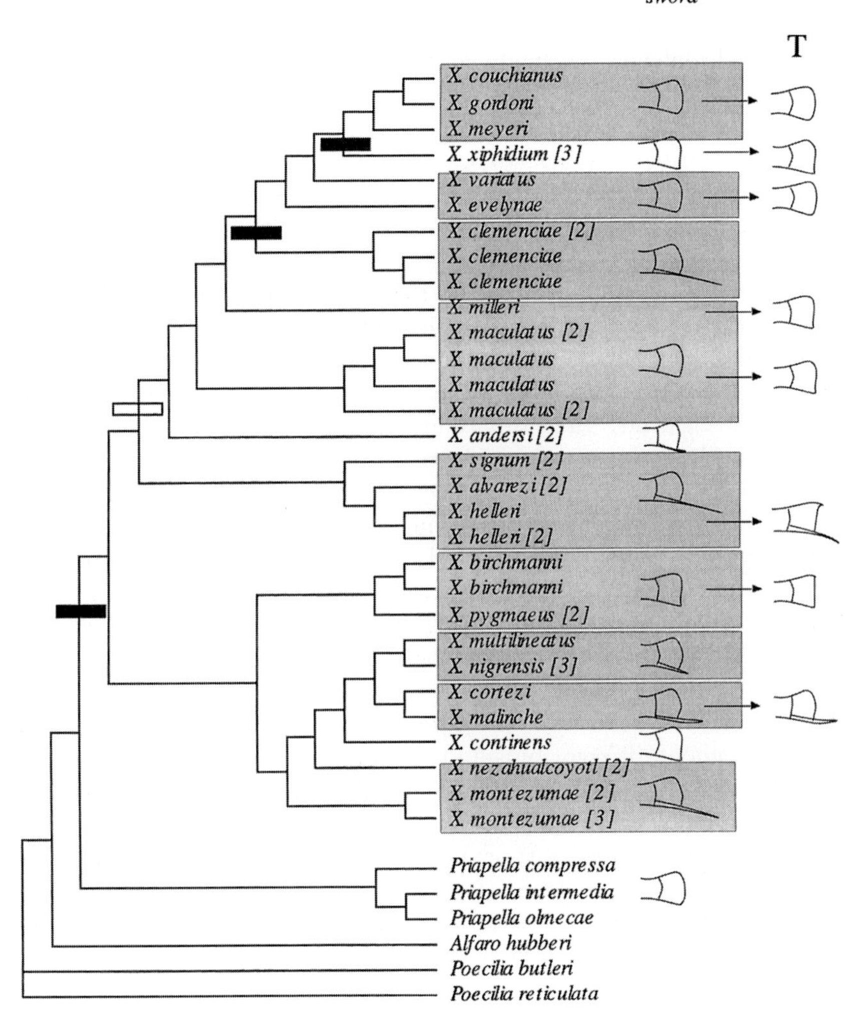

FIG. 4. Molecular (based on about 1300 bp of two mitochondrial genes and one nuclear gene) phylogeny of fishes of the genus *Xiphophorus*. The typical size of the ventral caudal fin extensions, in males of each species naturally and after treatment with testosterone, are indicated. Black bars indicate the presumed phylogenetic place of the origin, and the open bar indicates the loss of the sword within species of the genus *Xiphophorus*.

no longer expressed. It is conceivable that genes and gene networks might be retained by being used in several developmental contexts, e.g. the expression at different times, in different organs, or in the nexuses of different developmental networks. Most developmental control genes would seem to fit this description, and are expressed at various times during development and in different developmental contexts. This might create strong selection to maintain genes in their functional roles and in the genome. This retention of genes would then permit their recurrent co-option into new structures or networks, creating novelty, or similarly permit them to be 're-awakened' once they have been lost in the form of parallel or convergent morphologies. Novelty could arise in several ways: (1) retention and 're-awakening' of 'forgotten' genetic programmes possibly also through incorporation of this potentiality into a new context, as suggested here; (2) through gene duplications, by freeing one of the gene copies such that it may take on new functions (e.g. Ohno 1970); or (3) through regulatory evolution, by changing the control of expression of some genes or similarly by the co-option of genes into a new nexus of interactions among genes, through changes in regulatory elements. As more comparative developmental data become available, the relative importance of the retention compared to the evolution of novelty through gene duplication may become more apparent.

Developmental mechanisms are clearly important determinants of some macro-evolutionary phenomena, and they need to be incorporated into an extended modern synthesis of evolutionary biology. Only more comparative developmental data, analysed in a phylogenetic context, will allow us to better detect patterns in the diversification of life and to distinguish the major developmental mechanisms from those that are relatively unimportant. If homology describes the inevitable evolutionary phenomenon that the similarity of structures among different organisms is due to common descent—i.e. the information required for genetic and developmental mechanisms in evolutionary lineages is continuous—then analogy might be somewhat the same, except that the genes encoding analogous structures are not expressed continuously in all members of an evolutionary lineage, but are 're-awakened' during evolution.

The long retention of genetic systems—whether they are continuously expressed in all descendants of an evolutionary lineage in the form of homology, or only intermittently in the form of parallel evolution or even convergent evolution—may hence be caused by similar biological bases and may be due to similar evolutionary principles. Homology and non-homology might thus be extremes along a continuum rather than two completely different kinds of mechanisms.

Acknowledgements

Some of the research described in this essay is supported by grants from the National Science Foundation, the Max-Planck-Society and the University of Konstanz.

References

Abouheif E 1999 Establishing homology criteria for regulatory gene networks: prospects and challenges. In: Homology. Wiley, Chichester (Novartis Found Symp 222) p 207–225
Abouheif E, Akam M, Dickinson WJ et al 1997 Homology and developmental genes. Trends Genet 13:432–433
Basolo AL 1990 Female preference predates the evolution of the sword in swordtail fishes. Science 250:808–810
Basolo AL 1991 Male swords and female preferences: response. Science 253:1426–1427
Futuyma DJ 1998 Evolutionary biology, 3rd edn. Sinauer, Sunderland, MA
Gerhart J, Kirschner M 1997 Cells, embryos and evolution. Blackwell Scientific Publishers, Oxford
Halder G, Callaerts P, Gehring WJ 1995 Induction of ectopic eyes by targeted expression of the *eyeless* gene in *Drosophila*. Science 267:1788–1792
Hall BK (ed) 1994 Homology: the hierarchical basis of comparative biology. Academic Press, San Diego, CA
Jacob F 1977 Evolution and tinkering. Science 196:1161–1166
Kirschner M, Gerhart J 1998 Evolvability. Proc Natl Acad Sci USA 95:8420–8427
Lankaster ER 1870 On the use of the term homology in modern zoology, and the distinction between homogenetic and homoplastic agreements. Ann Mag Nat His 6:34–43
Meyer A 1993 Phylogenetic relationships and evolutionary processes in East African cichlids. Trends Ecol Evol 8:279–284
Meyer A 1997 The evolution of sexually selected traits in male swordtail fishes (*Xiphophorus*: Poeciliidae). Heredity 79:329–337
Meyer A, Kocher TD, Basasibwaki P, Wilson AC 1990 Monophyletic origin of Lake Victoria cichlid fishes suggested by mitochondrial DNA sequences. Nature 347:550–553
Meyer A, Morrissey JM, Schartl M 1994 Recurrent origin of a sexually selected trait in *Xiphophorus* fishes inferred from a molecular phylogeny. Nature 368:539–542
Nelson G 1994 Homology and systematics. In: Hall BK (ed) Homology: the hierarchical basis of comparative biology. Academic Press, San Diego, CA, p 101–149
Ohno S 1970 Evolution by gene duplication. Springer-Verlag, Heidelberg
Owen R 1848 On the archetype and homologies of the vertebrate skeleton. John van Voorst, London
Panchen AL 1994 Richard Owen and the concept of homology. In: Hall BK (ed) Homology: the hierarchical basis of comparative biology. Academic Press, San Diego, CA, p 21–62
Sanderson MJ, Hufford L (eds) 1996 Homoplasy: the recurrence of similarity in evolution. Academic Press, San Diego, CA
Sturmbauer C, Meyer A 1992 Genetic divergence, speciation and morphological stasis in a lineage of African cichlid fishes. Nature 358:578–581
Wake DB 1994 Comparative terminology. Science 265:268–269
Wake DB 1999 Homoplasy, homology and the problem of 'sameness' in biology. In: Homology. Wiley, Chichester (Novartis Found Symp 222) p 24–46
Zardoya R, Abouheif E, Meyer A 1996 Evolutionary analyses of *hedgehog* and *Hoxd-10* genes in species closely related to the zebrafish. Proc Natl Acad Sci USA 93:13036–13041

DISCUSSION

Tautz: How do we know that the genes were retained? Any gene in the genome acquires mutations continuously and can only be maintained if there is positive selection on it. Thus, genes that have no function would inevitably be lost from the genome, at least in the long run.

Meyer: It's probably a matter of the relative frequency of mutations compared to the strength of selection.

Wilkins: One explanation is the multiple developmental functions of genes and that it is possible to select for the retention of a gene by using it in a different developmental context.

Meyer: Yes, it may have multiple functions. Almost all genes are switched on in different places and at different times in development.

Roth: You mentioned the 'sword-making genes', but it remains to be shown that what was present in the ancestor and then retained is actually a mechanism for making a sword. If the mechanism has other functions, it may just be coincidental that it can produce a sword. To call it a sword-making apparatus may belie what its function is in lineages in which it has actively been maintained by selection without the sword being expressed. Is this conceivable?

Meyer: Females prefer to mate with males that have swords, and this female preference has been hypothesized to retain swords, even though there is a cost because males with swords are more visible to predators and less able to escape. Presumably, in some cases this cost outweighed the female preference benefit, and so swords disappeared.

Roth: Is it possible that the sword-making apparatus did not originate in conjunction with sexual display, or even with production of a sword on the tail?

Meyer: Yes, it's possible that it was co-opted for that function.

Roth: So the swords themselves may not be homologous, but rather the apparatus is homologous, in the sense of phylogenetic continuity.

Meyer: For argument's sake, if we assume that swords were lost and re-evolved twice, so that they evolved by parallel evolution since within the genus there were sword-less ancestors, are those swords not homologous to the swords that existed originally?

Wagner: A question in this context is whether the character was indeed lost. Since there are sword-less forms in this genus, the question is whether the ventral fin rays remain differentiated in the sword-less forms. If so, one would again face a question of character delineation. What is described as a loss and regain of sword may just be different states of a character that consists of the developmentally individualized ventral fin rays.

Meyer: Yes and no. In some of these species there is large amount of population variation. Both sword length and pigmentation have a normal distribution.

Rudolf Raff: Part of the problem is that you're looking at the sword as an isolated item, which isn't necessarily the case. It is likely that there are regionalized differences in the rays of the fin, for example. When you're talking about a sword gene, you're not necessarily talking about a gene that directly makes the fin, but rather a gene that affects hormonal control, or some other aspect of development, and it is this that has been selected for and against. In this sense, the entire process is homologous.

Meyer: I agree to some extent, but I don't necessarily see a difference in terms of whether selection acts on testosterone receptor genes or on other genes in the cascade of gene interactions that will eventually lead to a sword in an adult male fish.

Rudolf Raff: But it's a matter of whether you see these characters as arising in development or whether you see development as having numerous interacting components. If this is the case, then the interacting components can yield complex outcomes without disassembling the machinery at all, because what is important is how the elements in the machinery are interacting. In this case it may be a set of responder cells, and somewhere else in the body it may be something that regulates hormone levels.

Meyer: The reason I chose to talk about *Xiphophorus* was to illustrate how rapidly things can appear, disappear and reappear. A student in my lab is going to look at these sword genes by analysing cDNA libraries from regenerating cells of swords that have been cut off. This may or may not be the way to find these genes.

Müller: But you're not simply saying that things can disappear and reappear. You seemed to be saying that homoplasy and homology are more or less the same thing, and I'm surprised how you can reach this conclusion, because homoplasies arise independently in different lineages again and again by the retention of some developmental potential. Whether this is genetically determined or epigenetically determined is not important. The potential is there and so these characters arise. The switch to homology occurs when they become an integrated part of a bauplan that is the basis for further diversification of lineages, and the character is fixed within that bauplan. I would say that all homologues arose, at some point, as homoplasies, but they become homologues when they assume an organizational role in morphological evolution.

Meyer: In my opinion, a sword that reappears is the same as the original sword. It is still a homologous sword even though it disappeared for a short time in evolution.

Akam: So you would have no problem if it disappears every other generation in the female line. One could even imagine a continuum if it disappears in winter and comes back in summer, i.e. a seasonal polymorphism.

Maynard Smith: I would like to issue a slight warning here. We wrote a paper in *Nature* recently, not specifically on the problem we are discussing, but on the problem of control circuit redundancy and how two genes that do only one thing can be maintained (Nowak et al 1997). This is an important problem, and we had to work quite hard to think of mechanisms and explanations for the maintenance of genes against the constant noise of mutation. I would urge people who want, in any context, to invoke the explanation 'this genetic machinery has been there, although not expressed in a particular form, for quite a long time and it is now being recalled' to ask themselves what maintained it in the interval. In the case of the direct development of an echinoderm there probably isn't a problem, because almost all the genetic material needed to do the direct development of an echinoderm is presumably needed to turn the rudiment in a pluteus into an adult, so it has been maintained by selection because it is doing something anyway. I'm not trying to lay down the law. I'm just asking people to think carefully about the assumptions.

Meyer: My point is that there is so much parallelism that it may mean something. The sword genes, whatever their actual function is, have been maintained for 5–10 million years, and in the cichlids many life history traits and morphogenetic mechanisms have been maintained for even longer time periods. Another example is the evolution of giving birth to live animals in sharks, which re-evolved 10 times.

Galis: I'm puzzled by your statement that the possession of pharyngeal jaws is a specialized condition in cichlids because pharyngeal jaws occur in many fish families, and their possession in cichlids is certainly plesiomorphic. In cichlids only minor anatomical changes of the pharyngeal jaw apparatus are assumed to be responsible for the diversity (Galis & Drucker 1996).

Meyer: I am aware of that, but there are many other families of fish that have pharyngeal types of jaws of one kind or another, but they don't have this particular arrangement, as cichlid fish do. They are the only family in the African great lakes that do.

Abouheif: I have always been confused about the distinction in the literature between convergence and parallelism. Is it possible to distinguish clearly between the two?

Rudolf Raff: It depends on the mechanism. It is possible to have convergence without having similarity of mechanism or of parts, e.g. my leg and a table leg, whereas parallelism is assumed to involve the same mechanism recurring in two lineages to produce a similar effect. Obviously there are grey zones, but the concepts remain useful.

Wagner: A more primitive way to distinguish them is that parallelism is repeated evolution of a character starting from the same starting point, whereas convergent characteristics have different starting points.

Elizabeth Raff: There are many examples where different molecular pathways give rise to the same phenotypic result; thus, it would be difficult or indeed impossible to unravel whether the end result (the character) is convergent or parallel.

Carroll: The cladistic viewpoint is that there is no distinction between parallel or convergent evolution. It is a question of whether the character was or wasn't present in the immediate common ancestor, which is fairly easy to live with. I could never understand Simpson's marked distinction between parallelism and convergence; according to cladistic methodology, it's all homoplasy (Simpson 1953).

Nicholas Holland: During his career, Simpson made two distinctions. He started out with a strictly geometric model of parallelism and convergence, and then some years later he shifted into describing these terms in relation to genetic propensity.

Akam: The distinction is critical if you're concerned with mechanisms. Do you reiterate the same developmental changes in two parallel lineages to generate the same character transition, or do you do it by quite different methods? If you're thinking mechanistically this is important, but if you're thinking purely taxonomically perhaps it isn't.

Meyer: There are two grey zones: (1) does convergence start at the level of the genus or the family; and (2) how different is different, and how same is same, in terms of developmental mechanisms?

Maynard Smith: Isn't it just a question of going back to the latest common ancestor of the two groups you are comparing?

Elizabeth Raff: But you will never know what the mechanism was in the ancestor.

Wilkins: This is a probability argument. The further back you have to go to reach a common ancestor the less likely it is that the same genetic potential has been evoked. The problem is that it's difficult to put a number to that probability.

References

Galis F, Drucker EG 1996 Pharyngeal biting mechanics in centrarchid and cichlid fishes: insights into a key evolutionary innovation. J Evol Biol 9:641–670

Nowak MA, Boerlijst MC, Cooke J, Maynard Smith J 1997 Evolution of genetic redundancy. Nature 388:167–171

Simpson GG 1953 The major features of evolution. Columbia University Press, New York

Homology in the nervous system: of characters, embryology and levels of analysis

Georg F. Striedter

Department of Psychobiology and Center for the Neurobiology of Learning and Memory, University of California at Irvine, Irvine, CA 92697, USA

Abstract. The establishment of homologies is critically dependent upon the process of character identification. Valid characters must reliably appear in many individuals and be delimitable from other characters. They are not defined by any essential attributes, but rather by the formation of distinct clusters in a multidimensional morphospace. Features in two or more species can be considered possible homologues only if they are identifiable as the same character, for it would be nonsensical to homologize them as different characters. In order to confirm that a character is indeed homologous between species, one must examine its phylogenetic distribution to determine that it is unlikely to have evolved several times independently in the taxa being compared. This method of homologue identification can be applied to embryonic as well as adult characters and to characters at various levels of organization, including cell types and cellular aggregates. Difficulties arise, however, when one attempts to link the homology of adult characters to that of their embryonic precursors, or the homology of cellular aggregates to that of their constituent cell types. These efforts are misguided because different characters cannot be homologized to each other (as different characters). This perspective suggests that many neural characters may lack homologues, and therefore be truly novel, in other taxa.

1999 Homology. Wiley, Chichester (Novartis Foundation Symposium 222) p 158–172

The concept of homology is central to all of biology insofar as it is generally invoked, at least implicitly, whenever findings from one species are extrapolated to another (Wake 1994). So too in neurobiology, homology is widely used to justify research on so-called 'model systems' and to generalize across species (Striedter 1998a). Despite this widespread and often unambiguous usage, specific instances of homology are sometimes disputed so vehemently that doubts can be raised about the validity of homology as a concept (see Striedter 1998b). The principal thesis of this chapter is that most of the uncertainties surrounding homology result from difficulties with the first, but often neglected, step of establishing homologies, namely character identification. Although homologues in different species need not be identical, they must be identifiable as the 'same' character and attempts to homologize 'different' characters inevitably lead to confusion. Embryonic and adult features, for example, generally constitute entirely different characters, and it is misguided to infer the homology of adult characters solely from the homology of their embryonic precursors. Similarly, neuronal cell types and brain regions or nuclei form characters at different levels of analysis, and questions about their homology must therefore be logically uncoupled from one another. This analysis suggests that the concept of homology becomes logically coherent only when one acknowledges its limitations. Specifically, it suggests that many characters cannot be homologized across species and should therefore be regarded as genuinely 'new'.

Characters and the 'criteria' of homology

The question of what constitutes a valid character for homology comparisons is logically prior to the question of whether that character is homologous across species, but it is rarely explicitly addressed. Ghiselin (1966) was among the first to point out that any hypothesis of homology must specify what the features being compared are 'homologous as'. For example, Ghiselin argued that it is logically necessary to specify that bat wings and human arms are homologous as forelimbs. The word or phrase after the 'as' specifies the 'nature of the continuum' (Ghiselin 1966) or, in my interpretation, the identity of the character being homologized. Thus, bat wings and human arms can be identified as different characters with different names (wing versus arm), but they can also be brought together under the more inclusive character 'forelimb', which can then be homologized across humans and bats. Although it is probably best always to specify what the features being compared are 'homologous as', such specification is implicit when the features are identically named. 'The wings of birds are homologous' (Ghiselin 1966), for example, implies that the character being homologized is 'wing'. But how are a bird's wings initially identified as wings or forelimbs?

The process of character identification generally begins with an examination of numerous specimens within a species and, generally, across species. Comparative biologists then identify as valid characters those features that: (1) reliably appear in many individuals; and (2) are distinguishable from other characters within that same group of individuals. In essence, comparative biologists are trained to recognize which constellations of attributes (or parameters) recur consistently within a group of organisms or species (Striedter 1998b). Most importantly, characters are not defined by the possession of any essential attributes but rather by their formation of distinct clusters (or constellations) of attributes in multidimensional morphospace. Within the nervous system, for example, a neocortical area is recognized as such only if it is distinguishable from other cortical areas in terms of numerous attributes, including location, cytoarchitecture, physiology and function (Kaas 1983). If one of these attributes is lacking, e.g. visual input for a visual cortical area, then the cortical area may none the less retain its identity as that cortical area (Striedter 1998b). Similarly, retinal amacrine cells are identifiable as amacrine cells on the basis of numerous attributes, one of which is that amacrine cells typically lack an axon (amacrine literally means 'without axon'). However, some bona fide amacrine cells possess beautiful axons (Fig. 1; Dacey 1989).

If characters are not defined by any essential attributes, then homologues need not be identical in any particular aspect of 'form or function' (Owen 1848). However, one should also note that different characters can never be homologous

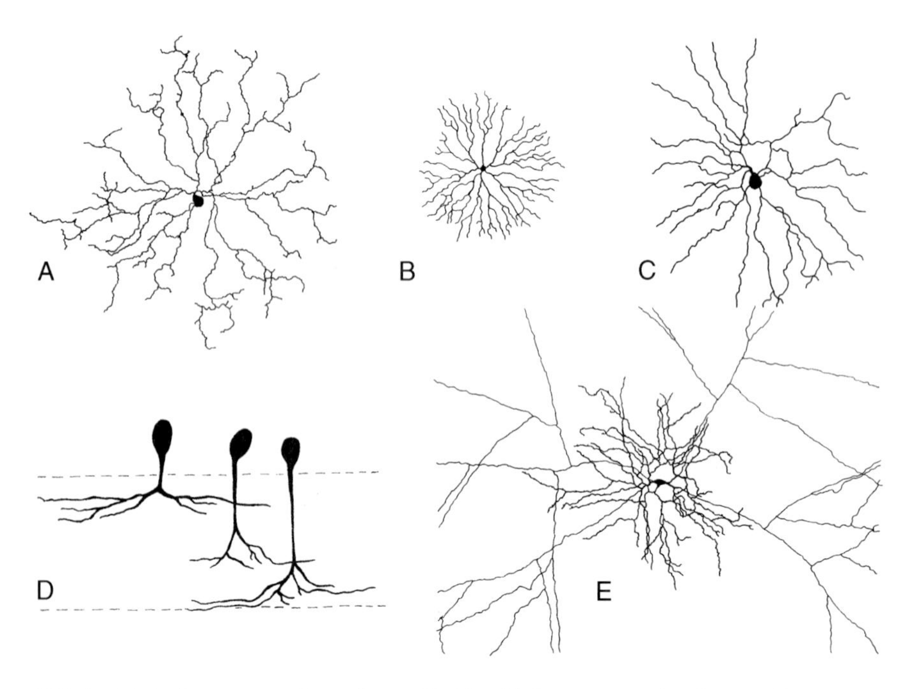

FIG. 1. Retinal amacrine cells can be readily identified as a neuronal cell type with numerous subtypes. Cells of the 'starburst amacrine' subtype have a characteristic dendritic branching pattern, best seen in horizontal sections through the retina (A–C). They are found in a broad variety of vertebrates, including humans (A), rabbits (B) and sharks (C). Transverse sections through the retina (D) reveal that amacrine cells in general extend their dendrites horizontally within the internal plexiform layer (between the dotted lines) and have their cell bodies located just outside this layer. Although amacrine cells typically lack axons, an axon-bearing subtype of amacrine cell has also been described (E). Redrawn from Cajal (1893), Rodieck (1989; reprinted by permission of John Wiley & Sons, Inc. Copyright © 1989), Dacey (1989; reprinted by permission of John Wiley & Sons, Inc. Copyright © 1989) and Brandon (1991).

to one another, for it would be illogical to suppose that two features can be homologous 'as' two different things. If one thinks of characters as clusters in multidimensional space, this means that disparate clusters cannot be homologous to one another unless they can be linked by intermediate forms (Remane 1952), in which case a more inclusive cluster (character) can be identified and be what the lower level characters are 'homologous as'. A lamprey cerebellum, for example, differs dramatically from a human cerebellum in terms of both structure and function, but numerous intermediate forms suggest that the character 'cerebellum' can none the less be recognized across vertebrates (Nieuwenhuys et al 1998). Similarly, the structure of pyramidal neurons differs dramatically between frogs and humans, but structurally intermediate forms in reptiles and mice permit the identification of the character 'pyramidal neuron' across tetrapods (Fig. 2; Cajal 1893, Nieuwenhuys et al 1998).

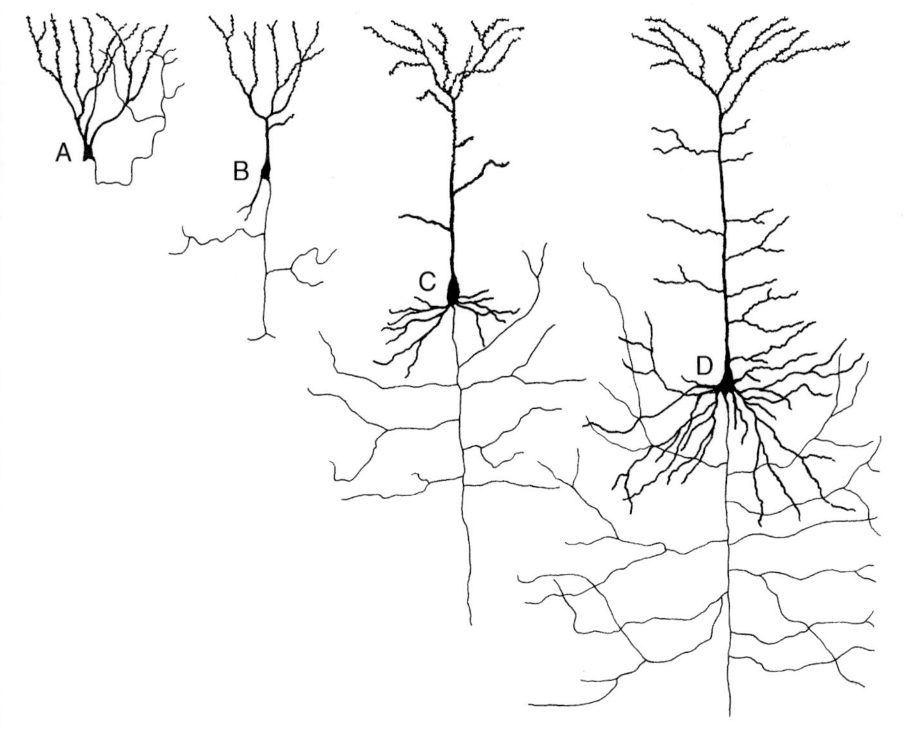

FIG. 2. Ramon y Cajal (1893) was the first to point out that neurons in the dorsal telencephalon of amphibians (A) can be homologized to pyramidal neurons in humans (D) through a series of intermediate forms seen in reptiles (B) and mice (C). A detailed discussion of how pyramidal neurons can be homologized across amniotes is found in Nieuwenhuys et al (1998). Redrawn from Cajal (1893).

Even if characters are 'the same' across species, however, they need not be homologous to each other, for the modern evolutionary definition of homology requires that homologous characters evolved only once, in a common ancestor of the species being compared, and were then retained with a continuous history in the descendant lineages (Striedter & Northcutt 1991). According to this definition, characters with multiple evolutionary origins must be due to convergent or parallel evolution, not homology. In order to apply this definition, information about the ancestral origins of characters must be inferred from the distribution of the characters among species whose phylogenetic interrelationships are known (Hennig 1966, Northcutt 1984). Thus, a character found only in distantly related taxa is probably not homologous between those taxa, whereas a character found in most members of a closely related group of species is probably homologous within that group (Striedter & Northcutt 1991). Organs capable of generating weak electric fields, for example, are found only in a few distantly related groups of teleost fish and are therefore unlikely to be homologous between those taxa as electrogenic organs (Bass 1986). The corpus callosum, on the other hand, is probably homologous among placental mammals because it is found in all placental mammals.

Some neural characters are not so easily compared across species, however, and their homologies are often debated endlessly. The homologies of cell groups in the telencephalon of birds and mammals, for example, have been debated for over 100 years, primarily because structurally intermediate forms are few and far between (Striedter 1997). Confronted with such divergent adult morphologies, different investigators have emphasized different kinds of similarity data to identify and homologize different telencephalic characters (Striedter 1997). Such disputes frequently give rise to debates about which of the so-called 'criteria of homology', e.g. neural connections or embryological origin, should be weighted most heavily in neural homology comparisons (Campbell & Hodos 1970, Nieuwenhuys et al 1998). Since characters are not defined by any essential attributes, however, and may change any of their attributes during the course of evolution, no single 'criterion of homology' can a priori be better than any other (Lauder 1994). In fact, the only essential criterion of homology is that the character has evolved once in a common ancestor and was then retained with a continuous history in the descendant lineages (Striedter & Northcutt 1991).

Therefore, prior debates about the relative merits of the various 'criteria of homology' are in actuality debates about 'criteria of character identification'. There are no generally optimal criteria of character identification, however, because different kinds of characters are best identified on the basis of different kinds of similarity data. Similarity in dendritic morphology, for example, is extremely useful in identifying specific neuronal cell types, but it hardly constitutes a useful 'criterion' in identifying major brain regions. This in turn suggests that the prior debates about the merits of different 'criteria of homology' have their origin in the tendency of different investigators to homologize different kinds of characters.

Embryonic characters and adult homology

By far the most frequently homologized neural characters are adult cell groups, such as the lateral geniculate nucleus, and major brain regions, such as the telencephalon. In order to identify and homologize such cellular aggregates, it is extremely useful to know where in the embryo they develop from, for experience has taught comparative neuroanatomists that cellular aggregates often conserve their site of embryonic origin even as their adult position, histochemical properties and neural connections diverge between species (Nieuwenhuys et al 1998). The utility of embryonic origin as a criterion for the identification of cellular aggregates was established through a *post hoc* analysis and therefore does not imply that homologous cell groups must always conserve their embryonic origin. It does, however, suggest that all vertebrate brains pass through a highly conserved embryonic stage, before and after which development diverges between species (Raff 1996). If so, then it should be relatively easy to homologize parts of the embryonic brain at this conserved stage of development. This was done explicitly by Bergquist & Källén (1954), who argued that the embryonic brain (at a stage soon after neural tube closure) is divisible into numerous distinct zones that can be identified in, and homologized across, all vertebrates. Although many aspects of this hypothesis remain controversial, recent studies have confirmed that at least some distinct zones exist in the embryonic nervous system and can be homologized across vertebrates (Lumsden 1990, Puelles & Rubenstein 1993, Nieuwenhuys et al 1998, Striedter et al 1998).

More controversial is the question of what the homology of such embryonic characters implies about the homology of their adult derivatives. This question is complicated partly because homologous embryonic characters often differentiate into different numbers of adult characters in different species, in which case one-for-one homologies will not be forthcoming for at least some of the adult characters (Bateson 1892). Some authors have attempted to solve this problem by invoking the concept of 'field homology', according to which adult structures derived from homologous embryonic precursors can be homologous 'as parts of a field' (Smith 1967, Butler 1994). The concept of field homology is vague, however, because it can be used to homologize almost any adult features, depending on how far back in embryonic time one is willing to search for homologous precursor regions (fields in the sense of Smith 1967). Field homology also cannot account for the observation that many patently homologous adult features, such as the vertebrate gut, develop from clearly non-homologous precursor regions in different species (see Striedter & Northcutt 1991). The concept of field homology therefore conflicts with what most evolutionary biologists mean by homology precisely because it establishes a rigid linkage between embryonic and adult homology (Striedter 1998b).

Instead of coupling embryonic homology to adult homology, it is more productive simply to homologize embryonic characters as embryonic characters and leave open the question of whether or not their adult derivatives are likewise homologous. Although few embryonic neural characters have thus far been identified and homologized, there are no conceptual obstacles to doing so. Embryonic characters can be identified as clusters in multidimensional morphospace on the basis of numerous attributes, such as patterns of cell division, migration, gene expression, lineage specification, relative position and adult fate. A standard phylogenetic analysis can then be used to determine whether the embryonic character is in fact homologous among the species being compared. The neural crest, for example, is a clearly delimitable embryonic character that is found in all craniates and, therefore, homologous among them. It is also clear that the neural crest contributes to a wide variety of adult tissues, e.g. teeth in larval amphibians and facial feathers in birds, not all of which are homologous to each other as adult characters (see Maderson 1987). One could argue that these adult characters are all homologous 'as derivatives of the neural crest', but this logically reduces to the statement that the neural crest is homologous 'as neural crest' and gives rise to a variety of adult characters in different species. It is clearer, therefore, simply to say that embryonic characters may be homologous as characters in their own right, and to acknowledge that such embryonic homology does not necessarily imply homology of the adult derivatives.

Although the homology of adult characters is conceptually independent from the homology of their embryonic precursors, embryonic characters may impart upon their adult derivatives some common attributes that make it possible to identify the collection of their adult derivatives as a single higher level adult character. The dorsal-most zone in the embryonic telencephalon of mammals, for example, develops into different numbers of cortical areas in different mammals, many of which cannot be homologized across species as individual cortical areas (Northcutt & Kaas 1995). However, all of the cortical areas derived from this dorsal embryonic zone can be identified and homologized as the higher level character 'neocortex' because they are more highly laminated than other telencephalic regions (and therefore cluster together in a distinct region of morphospace). The prominent lamination of neocortex probably derives from the fact that the dorsal-most zone of the embryonic telencephalon possesses a number of unique developmental features, such as the cortical plate, that are probably related to lamination in the adult neocortex (Sidman & Rakic 1982).

Even in this case, however, the homology of the embryonic dorsal telencephalic zone remains logically separate from the homology of adult neocortex. Therefore, the finding that some neocortical neurons derive from embryonic regions other than the dorsal telencephalic zone (Anderson et al 1997) does not negate the homology of neocortex (as neocortex).

Cell types and cellular aggregates

The preceding section showed that: (1) characters can be identified at various developmental ages; and (2) homology at one age need not imply homology at other ages. In addition, the discussion of neocortex and neocortical areas suggested that: (3) temporally synchronous characters can be identified at various levels of organization; and (4) homology at any one level need not imply homology at the other levels. The latter two points had been discussed previously in the context of general comments about the hierarchical organization of biological systems (Striedter & Northcutt 1991, Lauder 1994, Bolker & Raff 1996). None the less, the hierarchical nature of the homology concept remains potentially confusing because biological systems actually consist of multiple hierarchies that are interrelated in complex ways (Eldredge 1985). Within the nervous system, it is useful to distinguish at least three hierarchies, namely: (1) a hierarchy of cellular aggregates, composed of major brain regions, brain nuclei and nuclear subdivisions; (2) a hierarchy of cell types, including major types and subtypes; and (3) a hierarchy of molecules, grouped into families and superfamilies. Numerous homologies have been established within each of these hierarchies. Conceptual difficulties are encountered, however, when attempts are made to synthesize knowledge about homologies in different hierarchies. Most commonly, investigators encounter problems when they attempt to relate the homology of cell types to the homology of cellular aggregates (Reiner 1996).

These problems are not due to any conceptual difficulties in identifying or homologizing cell types *per se*. Just like cellular aggregates, cell types are identified by their formation of distinct clusters within a multidimensional morphospace (Tyner 1975, Harpring et al 1985). This morphospace for cell types is defined by attributes such as dendritic shape, axonal branching pattern, soma size, histochemical profile, intrinsic physiology, cell lineage, inputs, outputs and location within the nervous system. Although one need not examine all of these attributes to identify any particular cell type, attempts to identify cell types on the basis of just one or two attributes are bound to remain inconclusive. Furthermore, one should note that cell types are themselves hierarchically organized. Retinal amacrine cells, for example, can be identified and homologized as amacrine cells, but there are many different subtypes of amacrine cell, such as the starburst amacrines (Fig. 1), which can be identified as different characters at a lower level of analysis (Rodieck 1989, Brandon 1991). Once cell types have been identified as being the 'same' cell type in different species, their phylogenetic distribution must be examined to determine whether this 'sameness' is due to homology or independent evolution. The procedure for homologizing cell types is therefore precisely analogous to that for homologizing cellular aggregates.

But how is the homology (or non-homology) of cell types related to the homology (or non-homology) of the cellular aggregates in which they are located? It may be tempting to argue that cellular aggregates can be homologous only insofar as their constituent cells are likewise homologous, but this reductionist view is untenable because cellular aggregates are characters with their own emergent attributes (e.g. cytoarchitecture), not simply the sum of their constituent cells. If cellular aggregates and cell types are different characters, then cellular aggregates can be homologous (as aggregates) even if some of their constituent cells are not homologous (as cell types). The dorsal cortex of reptiles, for example, is most likely homologous to mammalian neocortex as a cellular aggregate despite the fact that reptilian dorsal cortex lacks many of the histochemically defined cell types characteristic of mammalian neocortex (Reiner 1991). A cellular aggregate can, therefore, during evolution gain or lose cell types without losing its identity as that cellular aggregate. In addition, it is possible for cellular aggregates to become lost during evolution even as their constituent cell types are retained. Teleost fish, for example, lack the deep cerebellar nuclei commonly found in other vertebrates but they possess cells within their cerebellar cortex that are probably homologous to cells in the deep cerebellar nuclei of other vertebrates (Fig. 3; Finger 1978, Murakami & Morita 1987). Homologous cell types, therefore, need not aggregate into homologous cell groups in different species.

A similar argument was made by Karten & Shimizu (1989) when they argued that the telencephalon of birds and mammals contains homologous cell populations which aggregate differently in the two taxa. Specifically, Karten and his collaborators identified similar intratelencephalic circuits in birds and mammals, hypothesized that equivalent nodes within these circuits could be homologized to each other as cell populations, and then argued that these homologous cell populations aggregate to form the laminar neocortex in mammals and the non-laminar dorsal ventricular ridge (DVR) in birds. Karten's cellular homologies were based almost exclusively on connectional similarities, and subsequent histochemical studies have shown that at least some of the cell types in mammalian neocortex cannot be identified in reptiles and birds (Reiner 1991). None the less, Karten's theory was logically sound insofar as it did not explicitly link the homology of cell types to the homology of cellular aggregates. Interestingly, however, Karten's theory was widely perceived to imply that avian DVR and mammalian neocortex must be homologous to each other as cellular

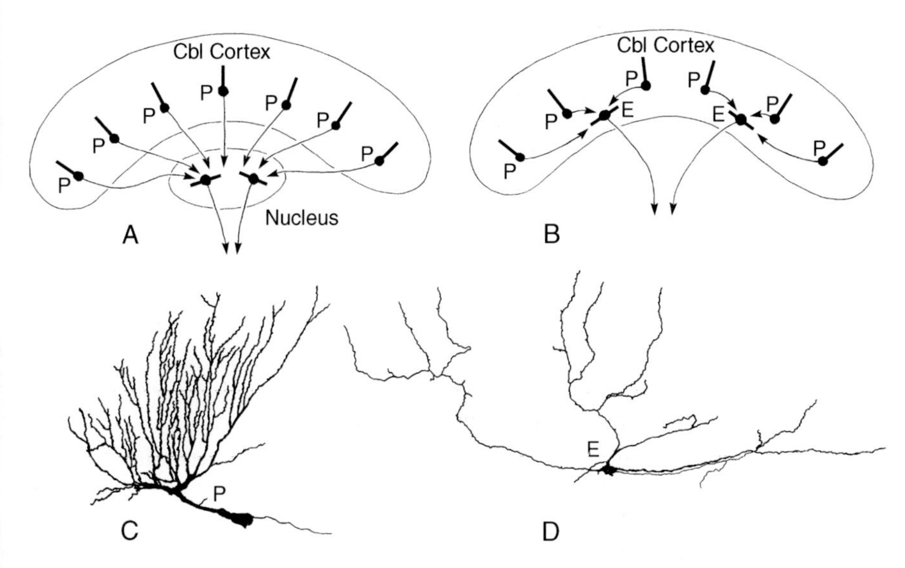

FIG. 3. Purkinje cells (P) in the cerebellar cortex of most vertebrates project to cells in a deep cerebellar (Cbl) nucleus, which then projects to targets outside the cerebellum (A). In teleost fish (B), however, Purkinje cells project to eurydendroid cells (E) in the cerebellar cortex, which then project to distant targets without synapsing in a deep cerebellar nucleus. These eurydendroid cells have been proposed to be homologous to the cells of the deep cerebellar nucleus in other vertebrates (as cerebellar efferent cells; Finger 1978). Golgi-stained Purkinje and eurydendroid cells in the cerebellar cortex of a trout are shown in (C) and (D), respectively (redrawn with permission from Pouwels 1978).

aggregates simply because their constituent cell populations are (hypothesized to be) homologous as cell populations (Northcutt & Kaas 1995, Reiner 1996, Striedter 1997). This argument is fallacious, however, because the homology of cellular aggregates does not stand or fall with the homology of their constituent cell populations. Even if some of the cell types in neocortex and DVR were homologous to each other (as cell types), neocortex and DVR would not necessarily be homologous to each other (as cellular aggregates). Conversely, one should not conclude that DVR and neocortex must be non-homologous (as cellular aggregates) simply because they contain some non-homologous cell populations (Reiner 1996).

Conclusion

The principal argument in this chapter is that features can be homologous to each other only if they are identifiable as the same character. Homologous characters need not be identical in every respect, but one cannot homologize different characters. It is makes little sense, for example, to homologize an embryonic precursor region to an adult structure, or a cell type to a cellular aggregate. This is not a radical view, for most comparative biologists have historically attempted only to homologize characters that are the same (Owen 1848). However, as embryological data become more widely available, and as comparative neuroanatomists look more closely at the cell types composing cellular aggregates, it will become increasingly important to state explicitly which characters are being homologized and then to refrain from attempts to homologize different characters.

One correlate of the above conclusion is that many adult characters may lack homologues in other taxa. The currently dominant view of brain evolution holds that most neural characters are conserved across species and that species differences arise primarily by the divergent modification of homologous characters (Butler & Hodos 1996). Although many neural characters clearly are conserved between species, an enormous number of neural characters cannot be identified (as the same characters) in all taxa. Comparative neurobiologists have sometimes attempted to homologize these apparently unique characters as: (1) derivatives of homologous embryonic precursor regions; or (2) components (or aggregates) of homologous characters at other levels of analysis. In doing so, however, they have implicitly changed the character that is being homologized. The apparently unique character therefore remains without a homologue in other species. Logically, then, it is better simply to acknowledge that many characters cannot be homologized, particularly among adult organisms, and that the evolutionary process continually creates truly novel characters by modifying the developmental fate of embryonic precursors and reshuffling characters at other levels of organization.

Acknowledgements

Many thanks to the organizers of this symposium for bringing together scientists from such diverse fields to debate a concept so central to all of biology. My own views on homology were influenced most by R.G. Northcutt, with whom I have extensively discussed many of the ideas presented here. Other colleagues with whom I have had the pleasure of discussing matters of homology over the years are Helmut Wicht, Wolfgang Plassman, Chris Braun, Harvey Karten, Anton Reiner and George Lauder. My research has been supported by the National Science Foundation and the National Institutes of Health.

References

Anderson SA, Eisenstat DD, Shi L, Rubenstein JLR 1997 Interneuron migration from basal forebrain to neocortex: dependence on *Dlx* genes. Science 278:474–476

Bass AH 1986 Electric organs revisited. In: Bullock TH, Heiligenberg W (eds) Electroreception. Wiley, New York, p 13–70

Bateson W 1892 On numerical variation in teeth, with a discussion of the conception of homology. Proc Zool Soc Lond 102–115

Bergquist H, Källén B 1954 Notes on the early histogenesis and morphogenesis of the central nervous system in vertebrates. J Comp Neurol 100:627–659
Bolker J, Raff RA 1996 Developmental genetics and traditional homology. BioEssays 18:489–494
Brandon C 1991 Cholinergic amacrine neurons of the dogfish retina. Vis Neurosci 6:553–562
Butler AB 1994 The evolution of the dorsal thalamus of jawed vertebrates, including mammals: cladistic analysis and a new hypothesis. Brain Res Rev 19:29–65
Butler AB, Hodos W 1996 Comparative vertebrate neuroanatomy: evolution and adaptation. Wiley-Liss, New York
Cajal SR 1893 Neue Darstellung vom histologischen Bau des Zentralnervensystems. Arch Anat Physiol Anat Abt Leipzig 29:161–181
Campbell CBG, Hodos W 1970 The concept of homology and the evolution of the nervous system. Brain Behav Evol 3:353–367
Dacey DM 1989 Axon-bearing amacrine cells of the macaque monkey retina. J Comp Neurol 284:275–293
Eldredge N 1985 Unfinished synthesis: biological hierarchies and modern evolutionary thought. Oxford University Press, Oxford
Finger TE 1978 Efferent neurons of the teleost cerebellum. Brain Res 153:608–614
Ghiselin MT 1966 An application of the theory of definitions to systematic principles. Syst Zool 15:127–130
Harpring JE, Pearson JC, Norris JR, Mann BL 1985 Subclassification of neurons in the ventrobasal complex of the dog: quantitative Golgi study using principal components analysis. J Comp Neurol 242:230–246
Hennig W 1966 Phylogenetic systematics. University of Illinois Press, Urbana, Il
Kaas JH 1983 What, if anything, is SI? Organization of first somatosensory area of cortex. Physiol Rev 63:206–231
Karten HJ, Shimizu T 1989 The origins of neocortex: connections and lamination as distinct events in evolution. J Cogn Neurosci 1:291–301
Lauder GV 1994 Homology, form, and function. In: Hall BK (ed) Homology: the hierarchical basis of comparative biology. Academic Press, San Diego, CA, p 151–196
Lumsden A 1990 The cellular basis of segmentation in the developing hindbrain. Trends Neurosci 13:329–335
Maderson PFA 1987 Developmental and evolutionary aspects of the neural crest. Wiley, New York
Murakami T, Morita Y 1987 Morphology and distribution of the projection neurons in the cerebellum in a teleost, *Sebasticus marmoratus*. J Comp Neurol 256:607–623
Nieuwenhuys R, ten Donkelaar HJ, Nicholson C 1998 The central nervous system of vertebrates. Springer, Berlin
Northcutt RG 1984 Evolution of the vertebrate central nervous system: patterns and processes. Amer Zool 24:701–716
Northcutt RG, Kaas JH 1995 The emergence and evolution of mammalian neocortex. Trends Neurosci 18:373–379
Owen R 1848 On the archetype and homologies of the vertebrate skeleton. John van Voorst, London
Pouwels E 1978 On the development of the cerebellum of the trout, *Salmo gairdneri*. Anat Embryol 153:37–54
Puelles L, Rubenstein JL 1993 Expression patterns of homeobox and other putative regulatory genes in the embryonic mouse forebrain suggest a neuromeric organization. Trends Neurosci 16:472–479
Raff RA 1996 The shape of life: genes, development and the evolution of animal form. University of Chicago Press, Chicago, IL

Reiner A 1991 A comparison of neurotransmitter-specific and neuropeptide-specific neuronal cell types present in the dorsal cortex in turtles with those present in the isocortex in mammals: implications for the evolution of isocortex. Brain Behav Evol 38:53–91
Reiner A 1996 Levels of organization and the evolution of isocortex. Trends Neurosci 19:89–93
Remane A 1952 Die Grundlagen des natürlichen Systems, der vergleichenden Anatomie und der Phylogenetik. Akad Verlagsges Geest & Portig, Leipzig
Rodieck RW 1989 Starburst amacrine cells of the primate retina. J Comp Neurol 285:18–37
Sidman RL, Rakic P 1982 Development of the human central nervous system. In: Haymaker W, Adams RD (eds) Histology and histopathology of the nervous system, vol 1. CC Thomas, Springfield, Il, p 3–145
Smith HM 1967 Biological similarities and homologies. Syst Zool 16:101–102
Striedter GF 1997 The telencephalon of tetrapods in evolution. Brain Behav Evol 49:179–213
Striedter GF 1998a Progress in the study of brain evolution: from speculative theories to testable hypotheses. Anat Rec 253:105–112
Striedter GF 1998b Stepping into the same river twice: homologues as recurring attractors in epigenetic landscapes. Brain Behav Evol 52:218–231
Striedter GF, Northcutt RG 1991 Biological hierarchies and the concept of homology. Brain Behav Evol 38:177–189
Striedter GF, Marchant TA, Beydler S 1998 The avian 'neostriatum' develops from the lateral pallium. J Neurosci 18:5839–5849
Tyner CF 1975 The naming of neurons: applications of taxonomic theory to the study of cellular populations. Brain Behav Evol 12:75–96
Wake D 1994 Comparative terminology. Science 265:268–269

DISCUSSION

Wagner: The importance of the limits to homology cannot be over-emphasized. Rieppel (1986) pointed out that the tendency of many comparative anatomists to a priori assume that there has to be homology among everything that can be compared is conceptually linked to the preformist idea that everything which exists must have existed before and nothing new can be generated. This is a non-evolutionary way of thinking. If we can understand what character identity actually means, we will also understand what innovation means.

Tautz: But even innovations must have had some previous structure on which they are based.

Wagner: But the products of these transformations are different from the precursors.

Rudolf Raff: This is a demonstration of features that we don't identify as homologues still having a continuity of information throughout development.

Wagner: Or rather a continuity of ontogenetic precursors because, for example, there are no structures, even in the sub-mammalian amniotes, that are in any way similar to the neocortex. Therefore, there is no continuity of information for the layering of the isocortex.

Rudolf Raff: But what makes evolution possible is that information, or if you like homologues at a lower level, are there to be reassembled into new things.

Wilkins: You are both correct because you are using the terminology in two different senses. Homology can mean characterization in terms of attributes, in which case bird and mammalian neocortices are not homologous; however, in terms of evolutionary origins bird and mammalian neocortices are homologous because they share an origin. The further lineages diverge in time, the less similar their attributes will become, so it will become more difficult to see what structures they share, and you need to take into account increasingly how they have come to differ.

Wagner: I agree that homology means two different things in this case. On the one hand, the identities of cells that end up in the isocortex are the same; and on the other hand, the result of phylogenetic derivation can be a new structure unlike the phylogenetic precursor. This means that the structure took on two different character identities in two different lineages.

Wilkins: So there is the descriptive meaning of homology and the historical meaning of homology.

Wagner: Or the mechanistic meaning of homology or character identity. It's not just descriptive.

Akam: It's easier for me to think of much simpler examples, e.g. what in a fly wing is the homologue of an eye spot in a butterfly wing? Clearly, some part of the fly wing is homologous to that part of the butterfly wing, but it doesn't make any sense to say that there is a homologue of a butterfly eye spot in a fly.

Müller: But if no character is individualized in the corresponding location of the organism used for comparison there is no point in trying to homologize something.

Hall: This is also the case for the digits of a tetrapod limb and a bony fish, which lacks digits.

Müller: But just because a character is not present in the primitive species does not mean we have to give up the concept of homology. Rather we should attempt to explain where the new homologue came from.

Nicholas Holland: Another way to look at this is to consider the situation in the pancreas. For example, the pancreas of higher vertebrates has β cells in the islets of Langerhans, but if you look at a larval lamprey you will find that the β cells occur in the submucosa of the gut wall. It is interesting in terms of evolution to look at the cellular level, because I can understand where the pancreas has come from during vertebrate evolution, but I don't call the lamprey gut wall the 'pancreas'.

Elizabeth Raff: Homology at one level need not imply homology at another level.

Hall: We are all in agreement that homology is hierarchical and it is not possible to homologize from one level to another.

Peter Holland: 'Not possible' is too strong. I would say that we shouldn't homologize from one level to another in all circumstances.

Roth: It still makes sense to homologize within a level, and then acknowledge any congruence among levels. The causal relationship that connects them doesn't justify homology across levels.

Peter Holland: That's right. The interesting biological question is, to what extent can we compare homologies across different levels, i.e. to what extent is there congruence?

Reference

Rieppel O 1986 Atomism, epigenetics, pre-formation and pre-existence: a clarification of terms and consequences. Biol J Linn Soc 28:331–341

Natural history and behavioural homology

Harry W. Greene[1]

Museum of Vertebrate Zoology and Department of Integrative Biology, University of California, Berkeley, CA 94720-3101, USA

Abstract. Although similarities and differences among animals have long inspired ethologists, three misconceptions haunt the literature on behavioural evolution: the absence of 'ethofossils' seriously hampers detection of behavioural homology; behaviour is more variable and subject to experiential modification than morphology; and behaviour is especially subject to convergence. As a backdrop to address these issues, I briefly survey parental care by amniotes and antipredator mechanisms in non-avian reptiles. Although those behaviours remain inadequately sampled taxonomically, they clearly vary at several cladistic levels; they are conservative across some major groups, innovative within subclades, and exhibit apparent homoplasy among and within groups. These behavioural examples also illustrate contextual influences on the expression of traits, as well as how behavioural context can shape other aspects of development. Enhanced understanding of behavioural evolution will follow from greater emphasis on how developmental context, including behaviour itself, shapes phenotypes; from integration of data for fossil and recent organisms; and from much denser ethological sampling among taxa. Phylogenetic analyses of behavioural similarity should in turn provide exciting insights into the evolutionary roles of behavioural shifts and constraints, as well as inform our aesthetic appreciation for the richness of nature.

1999 Homology. Wiley, Chichester (Novartis Foundation Symposium 222) p 173–188

'Most evolutionary biologists are interested in natural populations instead of fossils, and in variation among individuals rather than among species and higher taxa . . .'

(Brown 1982)

'But the muse of history says . . . when the future looks back on Darwin's age . . . it will see that . . . pre-evolutionary thought . . . was at last overcome by force of history — of change questions and tree thinking — and Clio came down from the rafters of our museums, shook off the dust, and took her rightful place in the director's chair. A happy outcome . . . (was that) when the Darwinian revolution came to a close at the end of the twentieth century, natural *history* became a discipline once again.'

(O'Hara 1988)

---

[1]Present address: Section of Ecology and Systematics, Corson Hall, Cornell University, Ithaca, NY 14853-2701, USA.

More than a dozen publications within the last few decades have directly addressed behavioural homology (e.g. Atz 1970, Brooks & McLennan 1991, Foster et al 1996, Greene 1994, Lauder 1986, Proctor 1996, Striedter & Northcutt 1991, Wenzel 1992, and numerous references therein), and especially with the recent appearance of another book on homology (Hall 1994), we might reasonably ask if there is anything more worth saying on this topic. Ironically, a question that has plagued evolutionary biology and seems to have inspired both this and Hall's earlier volume—what exactly does homology *mean*—has generally not been an issue with respect to behaviour. Since the inception of ethology as a discipline, its prevalent concern with homology has been detection of historical continuity, with interpreting similar behaviour across taxa as indicative of its presence in their common ancestor; controversy revolved instead around whether behaviour differed from morphology in ways such that one could not usefully apply that particular concept of homology to what animals do (see especially discussions in Brooks & McLennan 1991, Greene 1994, Wenzel 1992).

Here I will first describe phylogenetic patterns for parental care in amniotes and antipredator mechanisms in non-avian reptiles, then use those examples as the backdrop for addressing some misconceptions and a certain narrowness of vision that have haunted literature on behavioural homology. Then I will play off O'Hara's theme, embodied in the above quote, and argue that integration of historical perspectives with the details of natural history should play an increasingly prominent role in the study of behavioural evolution, as well as the aesthetic appreciation of biodiversity.

Continuity, sameness and divergence: two examples

Phylogenetic patterns of parental care among amniotes illustrate variation at deep, moderate and shallow levels of history (see reviews in Greene 1997, McKitrick 1992, Shine 1988). Some form of postnatal parental attendance characterizes virtually all mammals, crocodilians and birds, whereas that behaviour is relatively rare among other reptiles. Note that although fossil evidence for extinct archosaur behaviour (e.g. Carpenter et al 1994) is a welcome addition to the vertebrate ethogram, behavioural homology predicts the much ballyhooed social life of dinosaurs—in the face of its ubiquity in crocodilians and birds, their immediate living outgroups, what would be truly noteworthy is evidence that dinosaurs lacked parental care! Among squamates, parental attendance of eggs and/or young is reported for various scelroglossans, including some skinks (Scincidae), some alligator and glass lizards (Anguidae), pythons, mud snakes (*Farancia abacura*) and a few other species of colubrids, a few species of cobras and their relatives (Elapidae, even variably present among females shield-nosed cobras [*Aspidelaps scutatus*]), and many pitvipers (Crotalinae); parental care is lacking among iguanians (iguanas, chameleons, etc.), the sister taxon of Scelroglossa. Complex patterns of homologous and homoplastic parental care also are evident among living amphibians, the immediate outgroup to amniotes (Crump 1996), raising the possibility that absence of parental care in most turtles (but see Iverson 1990) and many squamates is derived. The ramifications of contact among parents and offspring (e.g. for incest avoidance, temperature preferences, feeding behaviour) might be substantial, but except for mammals and birds, we have scarcely begun to investigate those topics; likewise, we are almost totally ignorant of the behavioural processes underlying parental care in non-avian reptiles.

Historical evolutionary patterns of antipredator mechanisms among non-avian reptiles are distinctly hierarchical (reviewed by Greene 1988, 1997). Most species exhibit what is likely a retained amniote or even tetrapod set of responses, in which an animal reacts successively to threats by crypsis and immobility, locomotor escape, some form of non-combative defence (e.g. bluffs, distraction displays), and finally active fighting; struggling, biting and voiding cloacal contents also are widespread if not quite ubiquitous. Except for the shell itself, antipredator mechanisms of turtles are especially poorly studied, yet fossils illustrate armaments unknown among living species, the horns and a sort of mace-and-chain tail in extinct maiolanids.

Some defensive response has been described for about 10% of the ca. 2500 species of living snakes, making this group among the best studied of all amniotes in terms of antipredator mechanisms. Here again there is deep history, with primitive amniote patterns present in all basal snake groups (e.g. body thrashing, evacuation of cloacal glands), novel tactics characterizing various major clades and subclades (e.g. hoods in some elapids and xenodontine colubrids, tail vibration in pitvipers), present or absent among species (e.g. tail displays in *Micrurus*), and even variable among and within populations (e.g. tail displays of *Diadophis punctatus*). Homoplasy is seemingly extensive in defensive behaviour, with phylogenetic evidence for the independent evolution of tail displays, hoods, scale rubbing sound displays, open-mouth threat displays and aposematic colour patterns; however, some variation might be due only to threshold differences and thus represent undetected similarities among taxa (e.g. local variation in temperament of rattlesnakes, open-mouth threats in pitvipers). The fossil record of serpentine defensive displays is restricted to indirect morphological evidence of defensive tail displays in sand boas and their relatives (Erycinae). Macroevolutionary patterns are evident, e.g. open-mouth displays occur only within Colubroidea, the same subclade in which liberation of the maxillae from a major role in swallowing makes possible the origin of venoms and diverse other dental modifications for particular prey.

Towards consensus

The primary goal of behavioural comparisons often has been a search for additional 'taxonomic' characters. Perhaps that narrow conceptual focus, and the fact that most systematists work with morphology and molecules instead of behaviour, explain in part the relatively slow incorporation of behaviour in the phylogenetic revolution of the past decades (see especially Brooks & McLennan 1991). Three widespread presumptions also have stifled the historical comparative study of behaviour until recently: that absence of 'ethofossils' seriously hampers testing behavioural homology; that behaviour is more evanescent, variable and experientially modifiable than morphology; and that behaviour is under intense selection and thus rampantly convergent across taxa. I still hear similar sentiments expressed informally, almost three decades after the highly critical assessment of behavioural homology by Atz (1970) was published.

Each of these misgivings apply as well to morphology (Brooks & McLennan 1991, Greene 1994, Wenzel 1992), and the impact of missing fossils, variability, developmental malleability and homoplasy should be investigated rather than a priori used to exclude behaviour from phylogenetic analyses. Granting the existence of some fossil evidence for behaviours, most phenotypic attributes of most extinct organisms will always be unavailable for direct study, e.g. with rare exceptions (e.g. Haugh & Bell 1980) viscera don't fossilize, yet we have long relied on phylogenetic comparisons to homologize the four-chambered hearts among mammals, and the paired hemipenes that typify only lizards and snakes. As for behavioural evanescence, variability and modifiability, the bones that morphologists actually measure are usually no more than inanimate, desiccated shadows of what once were, at the tissue and molecular levels, highly dynamic and constantly changing living systems. And if transitory aspects of the phenotype are automatically suspect, are we not to compare across taxa the antlers of cervids or the uterine mucosa of mammals with oestrous? Some behaviour does exhibit extensive convergence among and within some groups, although the extent to which high apparent homoplasy might be due to failure to evoke threshold-dependent character states has scarcely been studied.

A fairly extensive literature and my brief examples show that behavioural characteristics can vary or remain stable at cladistic levels ranging from within and among populations to among and across ancient and highly distinctive lineages. Echoing David Wake's theme earlier in this volume, I urge that we get on with developmental and phylogenetic studies of behaviour, and in the next section I comment on prospects for that research. At the heart of the issues addressed by this symposium, we deal with continuity across taxa, and we face the loss of many species before they can be studied, as well as an amazing lack of detailed natural history for most organisms. Accordingly, I cannot resist closing with two of the same themes I raised in the volume by Hall (1994).

Challenges for the future

(1) As with other aspects of the phenotype, particularly exciting problems for evolutionary research concern the reality and causation of behavioural characters—not just what we see (for discussion of behavioural description, see Drummond 1981, Miller 1988), but also the processes underlying development and expression. In particular, the natural history of behavioural variations implies that homoplasy might be less common that implied by parsimony analyses, if all that really re-evolves is the expression of ancient traits. This problem must be addressed by strategic, detailed research on behavioural developmental mechanisms, as well as by field and naturalistic captive studies of the normal context for behavioural development and expression.

(2) The potential role of behaviour itself in shaping the development of phenotypes, including morphological, physiological and behavioural characters, deserves much more attention. Examples of behavioural influences on ontogeny include the effects of grooming behaviour and food manipulation on skull development in large cats (Duckler 1998), of nest temperature on sex determination in some amniotes (Ewert et al 1994), of prey size on head shape and size in watersnakes (Queral-Regil & King 1998), and of maternal care on brain function and subsequent responses to stress in rats (Liu et al 1997). We need to investigate the natural history of development—of the differential ontogeny in nature (or naturalistic settings) imposed by variation in oviposition sites, parental manipulation of eggs and young, proximity of siblings, and so forth.

(3) Enhanced understanding of behavioural homology will follow from much more thorough integration of data for extinct and extant taxa. On the one hand, palaeontologists have been slow to acknowledge the window that studies of modern behaviour provide for extinct taxa (but see papers in Thomason 1995); the discovery of Upper Cretaceous turtle eggs in dinosaur nests (Mohabey 1998), for example, is much more exciting in the context of living turtles that oviposit in alligator nests (Hall & Meier 1993). On the other hand, too often neontologists ignore relevant information from the fossil record. Neither of two recent papers on life history evolution in birds (Owens & Bennett 1995) and crocodilians (Thorbjarnarson 1996) include data on the other taxon or on dinosaur eggs (Carpenter et al 1994) as a basis for outgroup comparisons. Although the behavioural fossil record is sparse and its interpretation problematic (Greene 1994), palaeontology provides us with descriptions of predatory chases of

terrestrial invertebrates by pelycosaurs (Kramer et al 1995), diet and amount of mastication in giant theropods (Chin et al 1998), birthing behaviour in ichthyosaurs (Böttcher 1990), underground denning in early carnivorous mammals (Hunt et al 1983) and many other fascinating ancient events. My message here is that those studying behavioural evolution will profit from detailed knowledge of particular groups of organisms, regardless of whether the evidence comes from extinct or extant taxa.

(4) Complexities in the application of homology will inspire ever closer scrutiny with respect to mental states and their continuity among humans and other primates. Often politically charged, this topic ironically juxtaposes evolutionary parsimony and cognitive parsimony (Waal 1997): the former implies that the simplest and preferable explanation for shared similar behaviour in closely related species is common ancestry, whereas the latter (known as Morgan's Canon) demands the simplest available behavioural explanation (e.g. a conditioned response) for animals other than humans. Of course comparisons among neural structures have long been used for inferring and homologizing mental and sensory capabilities (even among extinct animals, e.g. Radinsky 1975), but now brain imaging and other new techniques raise the prospect of homologizing emotional, cognitive and motivational processes across taxa (Burghardt 1995, Wilson 1998). Clearly, issues of continuity and sameness in mental states will soon be enjoined with a directness not even imaginable a few decades ago.

(5) We should seek a general theory of behavioural macroevolution, of the tempo, mode and ramifications of historical stasis and change in behaviour across diverse groups of organisms. At the least, phylogenetic analyses of behavioural similarities and differences will provide exciting insights into the evolutionary roles of behavioural shifts and constraints with respect to morphological novelties and the radiation of higher taxa, e.g. the early evolution of tetrapods (Daeschler & Shubin 1995), mating behaviour and flightlessness in insects (Andersen 1997), and cranial kinesis in snakes (Greene 1994). New statistical approaches hold great promise for addressing rates of behavioural and morphological change (Gittleman et al 1996). More broadly, Alcock's tentative scheme for the evolutionary lability and stability of behaviour deserves detailed conceptual and empirical exploration (Alcock 1975; see Table 1): why should certain types of behaviour be resistant to change, and does the growing body of case histories for the age of particular behavioural homologies accord with those expectations?

(6) We must reduce the extreme shortage of behavioural descriptions for most groups. The reasons for the present lack of dense ethological sampling, beyond the sheer immensity of nature and small number of natural historians, are likely to include the incompleteness and limited value of much past descriptive work, the denigration in recent decades of non-question-oriented science and organismal

TABLE 1 A simple and tentative model of the tempo of behavioural evolution, adapted from Alcock (1975) and intended by him as an heuristic scheme for further development[a]

Behavioural category	*Highly labile*	*Moderately labile*	*Conservative*
Reproduction			
Courtship	✓		
Nesting		✓	✓
Agonistic		✓	✓
Grooming			✓
Habitat selection		✓	✓
Feeding			
Prey capture	✓	✓	✓
Food handling		✓	✓
Antipredator	✓	✓	

[a]Alcock reasoned, for example, that grooming actions are unlikely to facilitate competitive interactions, antipredator success or the invasion of new niches, and thus they should remain stable over long periods of evolutionary time; he noted that virtually all amniotes scratch their heads by lifting a hindlimb over a forelimb. Conversely, Alcock expected that courtship behaviour would be under severe selection for species specificity, and thus be especially labile.

biology in particular, and the chronic lack of funding for organismal and environmental biology. Encouraging taxonomically broad behavioural sampling would entail a stance rather different from the now prevalent 'model system' approach to science, since the former strategy would value stored inventories of basic knowledge for the sake of eventual applications. In terms of justification, a commitment to stock-piling descriptive studies would parallel the sequencing of entire genomes to serve various future goals, so our task is to explain more convincingly the important applications of natural history.

(7) Lastly, issues of behavioural sameness, continuity and divergence even apply to much broader societal problems, of which I stress the tragic and ongoing loss of biodiversity on a global scale. Kiester (1997) has framed the aesthetics of biological diversity with Immanuel Kant's distinction between beauty and the sublime; beauty characterizes objects that have attributes (e.g. colour, symmetry), whereas the sublime applies to that which is 'formless and unbounded and that strike(s) the imagination in a particularly powerful way.' One point in Kiester's extraordinary essay is especially relevant here: his concern that recent sophistication in systematic methodologies could overshadow individual taxa and the inherent differences among them—and thus detract from a sublime aesthetic attribute of biodiversity. I believe instead that phylogenetic perspectives (including phylogenetic taxonomy) provide inspiring temporal continuity to the similarities

and differences among organisms — that with knowledge of the evolutionary history of birds, for example, we might well care more about crocodilians.

Kiester's insight explains how historical perspectives on sameness and the divergence in natural history can enhance our sublime aesthetic appreciation of biodiversity, and thereby bolster efforts to conserve life on earth. The potential power of that approach for environmental education was emphasized for me this past spring, standing with my undergraduate herpetology class beside a small stream and surrounded by magnificent, primeval-feeling, California old-growth forest. We had just found a tailed frog (*Ascaphus truei*), so-named for the male's cloacal copulatory appendage, and my students were already aware that this single species is the sister taxon of all other living frogs (totalling ca. 4100 species). In short order I heard the following comments: 'Can you believe this creature hasn't shared its genes with any other frogs in more than 150 million years?', 'Yeah, but somehow it managed to invent this crazy mating tool found in no other amphibians . . .' And then, almost reverentially, 'Oh wow, this is it, we are looking at the basal frog!'

Acknowledgements

I thank G. M. Burghardt, B. K. Hall, G. B. Rabb, K. Schwenk and D. B. Wake for their insights; C. J. Bell for tutoring me in the richness of fossilized biology; and K. R. Zamudio for critical comments on the manuscript. The Lichen Foundation and the Museum of Vertebrate Zoology have most recently supported my field research on snake behaviour, much of it conducted in collaboration with D. L. Hardy Sr.

References

Alcock J 1975 Animal behavior: an evolutionary approach. 1st edn. Sinauer Associates, Sunderland, MA

Andersen NM 1997 Phylogenetic tests of evolutionary scenarios: the evolution of flightlessness and wing polymorphism in insects. Mem Mus Natl Hist Nat 173:91–108

Atz JW 1970 The application of the idea of homology to behavior. In: Aronson LR, Tobach E, Lehrman DS, Rosenblatt JS (eds) Development and evolution of behavior. WH Freeman, San Francisco, p 53–74

Böttcher R 1990 Neue Erkenntnisse über die Fortpflanzungs-biologie der Ichthyosaurier (Reptilia). Stuttg Beitr Natkd Ser B (Geol Palaeontol) 164:1–51

Brooks DR, McLennan DA 1991 Phylogeny, ecology, and behavior: a research program in comparative biology. University of Chicago Press, Chicago, IL

Brown JL 1982 The adaptationist program. Science 217:884–886

Burghardt GM 1995 Brain imaging, ethology, and the non-human mind. Behav Brain Sci 18:339–340

Carpenter K, Hirsch KF, Horner JR 1994 Dinosaur eggs and their babies. Cambridge University Press, Cambridge

Chin K, Tokaryk TT, Erickson GM, Calk LC 1998 A king-sized theropod coprolite. Nature 393:680–682

Crump ML 1996 Parental care among the Amphibia. In: Rosenblatt JS, Snowdon CT (eds) Parental care: evolution, mechanisms, and adaptive significance. Academic Press, San Diego, CA, p 109–144

Daeschler EB, Shubin N 1995 Tetrapod origins. Paleobiology 21:404–409

Drummond H 1981 The nature and description of behaviour patterns. In: Bateson PPG, Klopfer P (eds) Perspectives in ethology, vol 4. Plenum Press, New York, p 1–33

Duckler GL 1998 An unusual osteological formation in the posterior skulls of captive tigers (*Panthera tigris*). Zoo Biol 17:135–142

Ewert MA, Jackson DR, Nelson CE 1994 Patterns of temperature-dependent sex determination in turtles. J Exp Zool 270:3–15

Foster SA, Cresko WA , Johnson KP, Tlusty MU, Willmott HE 1996 Patterns of homoplasy in behavioral evolution. In: Sanderson MJ, Hufford L (eds) Homoplasy: the recurrence of similarity in evolution. Academic Press, San Diego, CA, p 245–269

Gittleman JL, Anderson GC, Kat M, Luh H-K 1996 Comparisons of behavioural, morphological and life history traits. In: Martius E (ed) Phylogenies and the comparative method in animal behavior. Oxford University Press, Oxford, p 166–205

Greene HW 1988 Antipredator mechanisms in reptiles. In: Gans C, Huey RB (eds) Biology of the Reptilia, vol 16, Ecology B, Defense and life history. Alan R Liss, New York, p 1–152

Greene HW 1994 Homology and behavioral repertoires. In: Hall BK (ed) Homology: the hierarchical basis of comparative biology. Academic Press, San Diego, CA, p 369–391

Greene HW 1997 Snakes: the evolution of mystery in nature. University of California Press, Berkeley, CA

Hall BK (ed) 1994 Homology: the hierarchical basis of comparative biology. Academic Press, San Diego, CA

Hall PM, Meier AJ 1993 Reproduction and behavior of western mud snakes (*Farancia abacura*) in American alligator nests. Copeia 1993:219–222

Haugh BN, Bell BM 1980 Fossilized viscera in primitive echinoderms. Science 209:653–657

Hunt RMJ, Xiang-Xu X, Kaufan J 1983 Miocene burrows of extinct bear dogs: indication of early denning behavior of large mammalian carnivores. Science 221:364–366

Iverson JB 1990 Nesting and parental care in the mud turtle, *Kinosternon flavescens*. Can J Zool 68:230–233

Kiester AR 1997 Aesthetics of biological diversity. Hum Ecol Rev 3:151–157

Kramer JM, Ericson BR, Lockley MG, Hunt AP, Braddy SJ 1995 Pelycosaur predation in the Permian: evidence from *Laoporus* trackways from the Coconino Sandstone with description of a new species of *Permichnium*. New Mex Mus Nat Hist Sci Bull 6:245–249

Lauder GV 1986 Homology, analogy, and the evolution of behavior. In: Nitecki M, Kitchell J (eds) The evolution of animal behaviour. Oxford University Press, Oxford, p 9–40

Liu D, Diorio J, Tannenbaum B et al 1997 Maternal care, hippocampal glucocorticoid receptors, and hypothalamic-pituitary-adrenal responses to stress. Science 277:1659–1662

McKitrick MC 1992 Phylogenetic analysis of avian parental care. Auk 109:828–846

Miller EH 1988 Description of bird behavior for comparative purposes. Curr Ornithol 5:347–394

Mohabey DM 1998 Systematics of Upper Cretaceous dinosaur and chelonian eggshells. J Vertebr Paleontol 18:348–362

O'Hara RJ 1988 Homage to Clio, or, toward an historical philosophy for evolutionary biology. Syst Biol 37:142–155

Owens IPF, Bennett PM 1995 Ancient ecological diversification explains life-history variation among living birds. Proc R Soc Lond B Biol Sci 261:227–232

Proctor HC 1996 Behavioral characters and homoplasy: perception versus practice. In: Sanderson MJ, Hufford L (eds) Homoplasy: the recurrence of similarity in evolution. Academic Press, San Diego, CA, p 131–149

Queral-Regil A, King RB 1998 Evidence for phenotypic plasticity in snake body size and relative head dimensions in response to amount and size of prey. Copeia 1998:423–429
Radinsky L 1975 Primate brain evolution. Am Sci 63:656–663
Shine R 1988 Parental care in reptiles. In: Gans C, Huey RB (eds) Biology of the Reptilia, Vol 16, Ecology B: Defense and life history. Alan R Liss, New York, p 275–329
Striedter GF, Northcutt RG 1991 Biological hierarchies and the concept of homology. Brain Behav Evol 38:177–189
Thomason JJ (ed) 1995 Functional morphology in vertebrate paleontology. Cambridge University Press, Cambridge
Thorbjarnarson JB 1996 Reproductive characteristics of the order Crocodylia. Herpetologica 52:8–24
Waal FBMd 1997 Foreword. In: Mitchell RW, Thompson NS, Miles HL (eds) Anthropomorphism, anecdotes, and animals. State University of New York Press, Albany, NY, p xiii–xvii
Wenzel GW 1992 Behavioral homology and phylogeny. Ann Rev Ecol Syst 23:261–281
Wilson EO 1998 Consilience: the unity of knowledge. Alfred A. Knopf, New York

DISCUSSION

Maynard Smith: I'm fascinated by the origin of human language, which I regard as essentially the origin of the human species. Language depends on a specific 'language organ', and new organs don't come from nowhere; they usually come from pre-existing organs. We're moving into a period when we may be able to identify genes that have specific effects on language, and then find out the homologues of those genes in other animals. For example, what is a language gene homologue doing in a chimpanzee? This will provide us with clues about the origin of human language, which would tell us a lot about the origins of the human species.

Striedter: Deacon (1992) has looked at the homologues of Broca's area in monkeys. He has found that the Broca's area homologues do not seem to be required for vocal communication in monkeys. So this is probably a case of evolutionarily co-opting brain areas into the neural circuits mediating vocal communication.

Maynard Smith: But what do these areas do in other animals?

Striedter: At the moment, the functions are largely speculative. The monkey homologues of Broca's area probably process auditory information and have motor access to facial and laryngeal muscles.

Abouheif: In terms of character delineation at the level of behaviour, what sort of behaviours can be easily homologized? And would home range size of an animal be considered a trait that one could homologize across species?

Greene: This has been talked about most frequently with regard to motor patterns. There has been an evolution of terms from fixed action patterns to modal action patterns in the literature. We shouldn't exclude too many behaviours in advance. Home range size, for example, might reflect something that evolves, in terms of tendency to be active or movement patterns. You would probably hope to understand something a little more precise than just large home range versus small home range. The problems of character delineation for morphology is equally a challenging problem for behaviour — for example, there is an elegant article by Drummond (1981) that talks about domains of regularity in behaviour — yet there's almost no cross-talk between people studying this in behaviour versus morphology.

Wagner: One criterion for the developmental individuality of morphological characters is that their development can be triggered by some inductive signal in a new context, like *Drosophila* eyes induced by *Pax-6* expression in the wing (Halder et al 1995). Similarly, character delimitation in behavioural biology may be achieved by testing what set of actions can be triggered by a well-defined signal. These action patterns act like behavioural modules that are self-regulatory and independent of the context in which they occur, once they have been triggered. There is evidence that it is those action patterns which can be compared between species.

Greene: Certainly, for some kinds of behaviour there are vast domains of regularity within species, across species and across higher taxa.

Tautz: As far as I understand, the aim of the behavioural ecology research programme is to determine in certain ecological settings what sort of behavioural responses are common across species, i.e. to look for homoplasies rather than homologies. Therefore, one potential complication is that the homoplasy could be large.

Greene: There is no evidence that this is the case. People have compared levels of homoplasies across taxa for behavioural and morphological data sets and found that they are not significantly different. The two different research programmes — entailing search for homoplasy or homology — are aimed at addressing related but different questions, respectively: convergent evolution as predicted by an engineering approach and, more broadly, the history of behavioural change.

Maynard Smith: The behavioural ecology movement, which developed in the 1960s and 1970s, attempted to analyse the adaptive significance of selective forces operating on behaviour. This was, to a certain extent, a reaction against comparative anatomy, which we had been taught and had bored us stiff. We probably reacted a little too strongly against the fascination that people like Konrad Lorenz had in the homology of different behaviours. Not that I ever intellectually rejected the notion that it is possible to homologize behaviours between different animals, but rather I wanted to be involved with something more testable and analytical.

Abouheif: Does this mean we have to introduce the notion of constraints at the level of behaviour? Optimality theory has been quite successful in predicting the behaviour of organisms given a set of environmental conditions. Thus, behaviours that have been conserved over long periods of evolutionary time may be somehow constrained. Can this notion apply to behaviour?

Galis: In cases such as signalling, for example, there is often a lot of conservation, but for feeding there may not. The homologization of behaviour has long been popular because when people started constructing cladograms functional morphologists studying the function of structures wanted to be taken seriously, so they constructed cladograms of the motor patterns they studied. Many mistakes have been made by trying artificially to find similarities (Galis 1996). For example, in some analyses conditions have been over-simplified, so that the conditions are similar and the heritability values are over-estimated. I'm thinking of an article by Lauder (1983), in which he claimed that there was a completely new character trait in a molluscivorous centrarchid, but the motor pattern he found was characterized by strong activity in all the antagonistic and protagonistic muscles. This is always found in forceful activities, and it cannot be seen as an inherited trait that can be homologized.

Greene: The first edition of John Alcocks's behavioural textbook contained a table that listed predictions about what kinds of behaviour would be stable over evolutionary time and what kinds would vary (Alcock 1975). It is perhaps not surprising that this table disappeared. It is in my chapter in this book because it's like a heuristic seed of a theory of behavioural stasis and change. One would like to come up with a theory that could predict, on first principles, which kinds of behaviour should be stable and which should change, because it certainly doesn't exist.

Müller: Have you come across instances of atavistic behaviour in these snakes? Is there such a kind of behavioural potentiality?

Greene: I know of some defensive displays that appear to be lacking in some snakes, but if you threaten these snakes at lower temperatures they demonstrate an open-mouth threat display that has previously not been recorded in this taxa, although was present in close relatives. Therefore, this open-mouth threat display may have evolved many fewer types than one would conclude by just counting instances of expression on a cladogram.

Striedter: You mentioned that this open-mouth threat display in certain snakes evolved at the same time as venomous snakes. Is it possible that the venom-producing organs have been lost in these snakes? If so, it would be interesting to speculate that the open-mouth threat display had been retained, rather than it having evolved as a case of mimicry.

Greene: I don't know if that is the case. If you plot different defence displays on a more detailed cladogram of snake relationships, all the specialized defensive displays occur only at or above the node at which venom is characterized. Something happened in snake evolution that freed the maxillae from swallowing, and it is only after that point that the elaboration of maxillary tissue occurs, e.g. to form venom-injecting tissues.

Abouheif: Is there strong selection on the posture of the open-mouth threat display? Do you have a feeling that some snakes have a better posture than others?

Greene: I'm not sure about this trait. I've studied the application of constricting coils around prey, and it turns out that there are about 30 different ways of applying coils to prey. In some taxa this behaviour is extremely variable. Sometimes it is extremely variable at birth but then it becomes more stereotypical with repeated encounters of prey. However, in several basal lineages of snakes, e.g. boa constrictors and pythons, the very first time the snakes apply constricting coils to their prey, they do it in precisely the same fashion as the adult. Therefore, there are extreme levels of stereotypic behaviour. I tried to inject variability into the system—e.g. I tried making snakes constrict by hanging them from my finger instead of placing them on laboratory benches, and I tried making anacondas constrict in water to see if substrate could effect the topography—and I could not inject variability into this behaviour.

Rudolf Raff: Yet there are snakes that learn different hunting strategies.

Greene: Yes. I'm certainly not denying the possibility of modification by experience. For example, if you give baby snakes their first mouse and score which way the prey is swallowed, 99% of the time ingestion is head first. But if you watch carefully, you see that they try to swallow the hindfoot, the tail, the rear end, the whiskers; indeed, their first attempts are totally random. However, it is difficult to swallow a mouse any way except head first. They do demonstrate a learning curve though: with each successive experience with a mouse, they pick up more and more cues that they use to shorten the time on their next encounter.

Purnell: I would like to bring up the issue of the fossil record and behaviour. There are more than 600 million years of trace fossil evidence for behaviours, but has anyone tried to apply behavioural homology to this?

Greene: Not as much as they could. There are books on trace fossil elements of behaviour, i.e. ethofossils, and there are considerations of the possible biases of fossilized behaviour. It's likely that some fossilized behaviours will be fossilized because animals made fatal mistakes—they are thus fossils of abnormal behaviour—because why else would they be available for fossilization? In general, neontologists have not looked carefully at the palaeontological record. It will take people who are focused on the total literature of a particular group of organisms to look at this in detail.

Rudolf Raff: Seilacher (1986) tried to order the increasing complexities of behaviours in the lower Paleozoic fossil traces.

Greene: What you find much more often in the literature is someone studying extinct taxa, and, as detailed in my chapter, they are often ignorant of the biology of modern taxa, or vice versa.

Striedter: The relationship between behaviour and morphology is particularly interesting in terms of which comes first, the evolution of new morphologies or new behaviours? In my opinion we don't know enough to answer this question.

Müller: I was surprised how easily you dismissed the widespread tenet in textbooks of evolutionary biology that a change of function precedes a change of form. Could you comment further on this?

Greene: This tenet might be generally true, although there isn't a robust justification in the literature. We ought to be able to address this by looking at the distribution of shifts in behaviour and morphology — which appears first within a particular clade — but there are relatively few published analyses.

Galis: I disagree. Sometimes important behavioural changes will lead to morphological changes, and sometimes there will be morphological innovations that lead to behavioural changes.

Wilkins: It is also likely that these will occur in a stepwise fashion.

Striedter: Or you could say that they happen simultaneously, so that in that sense neither is first.

Carroll: There are numerous informative examples among secondarily aquatic reptiles. There are many clades that have gone back to the water, including animals such as the Galapagos marine iguana, which shows limited structural or physiological adaptations towards life in the water, and yet it is totally dependent on feeding on aquatic algae. In this case it is clear-cut that behaviours were there before structures. When we were working on the origins of hothosaurs and plesiosaurs we happened to have some green iguanas in the laboratory that had never been in the water. We placed one in a wading pool and watched as it tried to paddle for about 15 seconds, realized that this wasn't the way to swim, then swam in an anguilloform manner for longer periods than it could run on land. Therefore, the structure and physiology of these forms, because they're capable of withstanding cold and anoxia for a long time, are inherently adapted towards life in the water. Clearly, in these cases the behaviour comes first and then they gradually pick up the anatomy.

Galis: This is because they are occupying a new niche, and so plasticity is important. However, both behavioural plasticity and morphological plasticity are required.

Carroll: It takes a long time for them to change structurally. There is a good transition from terrestrial varanoids via an intermediate group to the mosasaurs, which were the dominant marine reptiles at the end of the Mesozoic period. What changed first is hearing, and this is also true in whales. The capacity to feed in the water was also achieved rapidly, but surprisingly the limbs are one of the last elements to become fully adapted to an aquatic way of life.

Galis: But in the case of the cichlids, structural changes precede behavioural changes.

Wagner: It is not clear what causes the structural innovations in cichlids. There are species that persist with the plesiomorphic phenotype, so it cannot be seen as a simple optimization step. It is likely that many of these key innovations are side-effects of other changes (Müller 1990). With respect to feeding, for instance, the structural innovations are caused by non-feeding adaptations, such as size of skull. In those cases a new function arises from a transformation that has nothing to do with this function.

Carroll: But changes in the middle ear of some secondarily aquatic vertebrates are specifically associated with underwater feeding and hearing, so it can operate both ways.

Greene: There must be someone who thinks that behaviour is irrelevant in itself as a context for development, i.e. that a wild cat's skull morphology is there in the genome, and that what I described for zoo animals is simply pathology.

Tautz: I do not see the connection between behaviour and development.

Greene: But if you change the experience of the organism you get a different phenotype.

Meyer: Animals that have larger, but fewer, eggs may have a different developmental pathway than those with smaller, but more numerous, eggs. This would be a behavioural shift that would have developmental consequences.

Abouheif: Another example of behaviour regulating phenotype is in ants. They regulate the population of the different castes in their colonies. If too many soldiers, for example, are produced in the colony, other soldiers feed the larvae appropriately so that they shift their developmental trajectory and become workers. Wheeler & Nijhout (1984) applied some juvenile hormone to switch the developmental trajectory of workers into soldiers. When they did this in the absence of soldiers, they continued to develop into soldiers, but when soldiers were present, the soldiers secreted some inhibitory pheromones and suppressed that developmental switch.

Hall: There are also examples of cyclomorphosis, in which a structure appears in the prey only in the presence of the predator.

Wray: And there is also the hard-wiring of stable switches in genetic programmes, such as seasonal polyphenisms in insects and sexual dimorphisms in many organisms. These are stable, distinctive states. They have their own ranges of variation that often don't overlap.

Müller: A human example is that the behaviour of drinking milk after weaning led to the development of a gene complex that prevents the suppression of the lactose enzyme in adult European populations, while this enzyme is not present in Asians or Africans.

Greene: It seems to me that the bulk of the discussion over the last two days has concentrated on sequences and molecular genetic control mechanisms, embryogenesis and adult forms, and what we're trying to discuss is similarities and differences in continuity in this adult form. The context for development seems to me to extend beyond the organism itself, and this external context seems to be an almost totally ignored aspect of developmental biology.

Rudolf Raff: Developmental biology has also ignored life history and phenotypic variation.

Abouheif: I would argue that in evolutionary biology the study of maternal effects has not received a lot of attention from molecular biologists. Quantitative geneticists are studying the effects of, for example, where a mother lays her eggs and the resultant effects, but no one is studying this at the molecular level.

Akam: But it's only recently that that most developmental biologists started to look beyond gastrulation. Until recently, the second half of embryogenesis was almost a black box. It is a field that has so far explored a few tiny islands and left continents of ignorance.

References

Alcock J 1975 Animal behavior: an evolutionary approach. 1st edn. Sinauer Associates, Sunderland, MA

Deacon TW 1992 Cortical connections of the inferior arcuate sulcus cortex in the macaque brain. Brain Res 573:8–26

Drummond H 1981 The nature and description of behaviour patterns. In: Bateson PPG, Klopfer P (eds) Perspectives in ethology, vol 4. Plenum Press, New York, p 1–33

Galis F 1996 The application of functional morphology to evolutionary studies. Trends Ecol Evol 11:124–129

Halder G, Callaerts P, Gehring WJ 1995 Induction of ectopic eyes by targeted expression of the *eyeless* gene in *Drosophila*. Science 267:1788–1792

Lauder GV 1983 Neuromuscular patterns and the origin of trophic specialization in fishes. Science 219:1235–1237

Müller GB 1990 Developmental mechanisms at the origin of morphological novelty: a side-effect hypothesis. In: Nitecki MH (ed) Evolutionary innovations. University of Chicago Press, Chicago, IL, p 99–130

Seilacher A 1986 Evolution of behaviour as expressed in marine trace fossils. In: Nitecki MH, Kitchell JA (eds). Evolution of animal behaviour, paleontological and field approaches. Oxford University Press, New York, p 62–87

Wheeler DE, Nijhout HF 1984 Soldier determination in *Pheidole bicarinata*: inhibition by adult soldiers. J Insect Phys 30:127–135

Evolutionary dissociations between homologous genes and homologous structures

Gregory A. Wray

Department of Ecology and Evolution, State University of New York, Stony Brook, NY 11794-5425, USA

Abstract. Phenotype is encoded in the genome in an indirect manner: each morphological structure is the product of many interacting genes, and most regulatory genes have several distinct developmental roles and phenotypic consequences. The lack of a simple and consistent relationship between homologous genes and structures has important implications for understanding correlations between evolutionary changes at different levels of biological organization. Data from a variety of organisms are beginning to provide intriguing glimpses of the complex evolutionary relationship between genotype and phenotype. Much attention has been devoted to remarkably conserved relationships between homologous genes and structures. However, there is increasing evidence that several kinds of evolutionary dissociations can evolve between genotype and phenotype, some of which are quite unexpected. The existence of these dissocations limits the degree to which it is possible make inferences about the homology of structures based solely on the expression of homologous genes.

1999 Homology. Wiley, Chichester (Novartis Foundation Symposium 222) p 189–206

The relationship between genotype and phenotype in multicellular organisms is complex and indirect. Although much of the nomenclature of genetics suggests a 1 : 1 relationship between a particular gene and specific phenotypic feature, a more realistic conceptual model is that intricate networks of interacting genes produce phenotypic outcomes (Fig. 1). Pleiotropy (several distinct phenotypic consequences from one gene) and polygeny (several genes contributing to one aspect of phenotype) are pervasive phenomena (Wright 1980, Falconer 1989, Wilkins 1993). Evolutionary dissociations are possible between homologous genes and homologous structures, as originally noted by de Beer (1971). This chapter reviews these evolutionary dissociations and their implications for making inferences about morphological homology based on the expression of homologous regulatory genes.

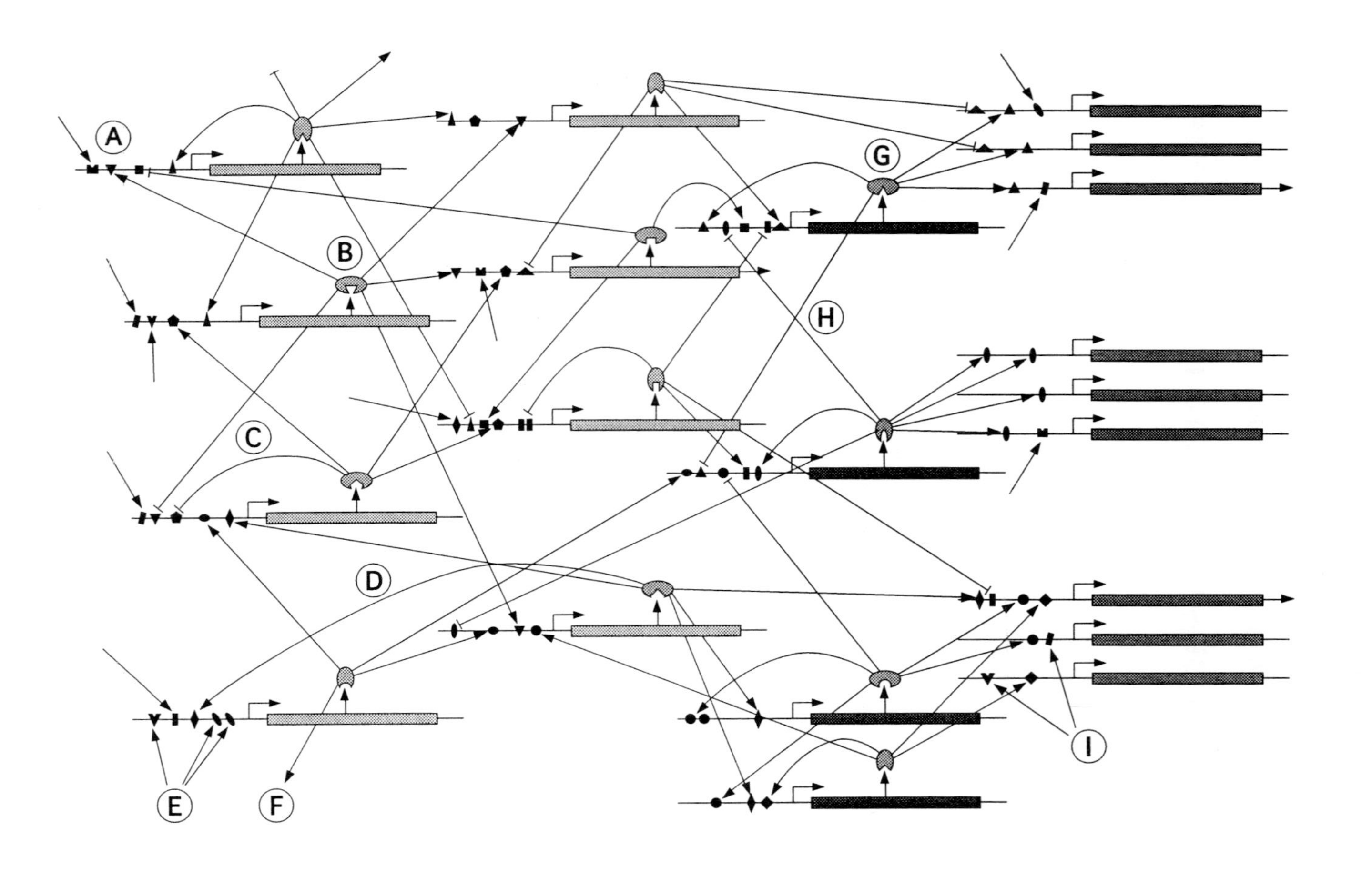

A
B
C
D
E
F
G
H
I

General properties of regulatory genes

Our understanding of the genes that regulate developmental processes is increasing exponentially. Although the primary motivation for studying these genes is elucidating developmental mechanisms, information about them is also relevant to understanding the evolution of morphology.

Regulatory genes typically perform several distinct developmental roles

The *engrailed* gene of *Drosophila* encodes a transcription factor that plays several distinct roles during development (Rogers & Kaufman 1996). Three lines of evidence suggest that this is a fairly typical situation. First, mutations in many regulatory genes have complex, pleiotropic phenotypes (Wilkins 1993). However, pleiotropy can arise in various indirect ways, and is not by itself strong evidence for multiple developmental roles. Second, most regulatory genes are expressed in several temporally and spatially distinct phases (Duboule 1994). Together, genetic and expression data can provide compelling evidence that a single gene is involved in several distinct developmental processes (Duboule & Wilkins 1998). By these criteria, most well-studied regulatory genes have multiple developmental roles. A third line of evidence for multiple roles is based on the observation that relatively few regulatory genes exist for each kind of developmental task. For instance, the receptor proteins that bind intercellular signalling molecules in animals belong to just a few different gene families. The number of times that specific cell signalling is needed during development vastly exceeds the number of signals and receptors present, necessitating multiple uses (for examples, see Bang et al 1995, Moskowitz & Rothman 1996).

Developmental roles evolve

Indirect evidence strongly suggests that regulatory genes can gain and lose developmental roles during the course of evolution. It seems highly implausible

---

FIG. 1. Hypothetical gene regulatory network. Interactions among regulatory genes are immensely complex. The most direct interactions within a network are between genes encoding transcription factors (shown here), but indirect interactions also occur through various intercellular signalling systems. The expression of any one gene is typically regulated by several different transcription factors (A; Arnone & Davidson 1997). Most regulatory genes, in turn, influence the expression or activity of several other genes or gene products (B; Duboule & Wilkins 1998). Regulatory networks contain strongly non-linear features such as autoregulation (C), indirect feedback (D), indirect inputs from processes outside the cell (E) and outputs to other processes (F). Some genes (G) regulate batteries of genes that produce differentiation products. Such 'master regulatory' genes may have mutually repressive interactions (H). Even genes that are activated by such genes can sometimes be activated in other ways (I). The degree to which regulatory interactions within a regulatory network persist or change during the course of evolution remains poorly understood, but examples of both long-term conservation and rapid change are known.

that the multiple developmental roles of a single gene arose simultaneously when the gene first appeared. Indeed, many regulatory genes in plants and animals belong to gene families that predate the origin of these two kingdoms (e.g. Bharathan et al 1997), and many of the developmental roles that these genes currently play in plants and animals are therefore unlikely to be their original ones. Comparisons of expression domains among taxa strongly reinforce this conclusion. Few (if any) of the multiple expression domains and developmental roles of *engrailed* in arthropods and chordates are homologous (Davis et al 1991, Duboule 1994, Rogers & Kaufman 1996). However, comparisons of regulatory genes among phyla sometimes do reveal overt similarities in expression domains. These similarities are usually interpreted as being homologous, rather than convergent (Slack et al 1993, Finklestein & Boncinelli 1994, De Robertis & Sasai 1996, Kimmel 1996). Even if this interpretation is correct for the particular cases that these authors discuss, it is none the less true that the majority of known developmental roles are apparently not conserved between animal phyla because they have no counterpart outside of one phylum. Comparisons within animal phyla provide many additional examples of probable evolutionary gains and losses of developmental roles (e.g. Patel et al 1992, Falciani et al 1996, Lowe & Wray 1997, Grbic & Strand 1998).

Interactions among regulatory genes are complex

Regulatory genes do not act in isolation, but as components in regulatory networks of extraordinary complexity. Although geneticists frequently use the term 'pathway' to describe interactions among genes, they include many non-linear features such as autoregulation, indirect feedback, environmental inputs and redundancy (Fig. 1). In addition, most regulatory genes appear to interact with many upstream and downstream genes. The promoters of eukaryotic genes often contain dozens of binding sites for many different regulatory proteins (Arnone & Davidson 1997). Regulatory genes may, in turn, have dozens of downstream target genes (e.g. Mastick et al 1995), although accurate estimates are difficult to obtain.

Interactions among regulatory genes can evolve

It has been clear for some time that regulatory interactions can evolve (Zuckerkandl 1994), but direct comparative analyses of interactions among regulatory genes are uncommon in the literature (for exceptions, see Shubin et al 1997 and Abouheif 1999, this volume). Some regulatory interactions are probably ancient, such as the role of Trithorax and Polycomb group genes in regulating *Hox* genes in arthropods and chordates (Schumacher & Magnuson 1997). These examples imply conservation of regulatory interactions in excess of half a billion years. On the other hand, regulatory interactions can also change over much shorter time-scales. Protein-binding sites within the *even-skipped* promoter differ among *Drosophila* species (Ludwig et al 1998), and this gene has both gained and lost a pair-rule segmentation role within holometabolous insects (Patel et al 1992, Grbic & Strand 1998). A variety of experiments provide direct evidence that regulatory interactions can change over relatively short evolutionary time-scales (Franks et al 1988, Klueg et al 1997, Meise et al 1998, Saccone et al 1998). More extensive, but indirect, evidence for evolutionary changes in interactions among regulatory genes comes from comparisons of expression domains among species, which reveal many examples of changes in the time, location and level of expression (Patel et al 1989, Davis et al 1991, Holland et al 1996, Savage & Shankland 1996, Lowe & Wray 1997, Grbic & Strand 1998).

Evolutionary association and dissociation between genes and structures

Homologous genes in two different species are sometimes expressed in homologous structures or regions of the body, implying a persistent evolutionary association between genes and structures (Fig. 2A). However, evolutionary dissociations of various kinds can occur between homologous genes and homologous structures (Fig. 2B–F). Before going on, three terms need to be defined: (1) 'homologous' is used here to denote a feature that is present in two species because it was present in their latest common ancestor, and genes belonging to families are considered (2) orthologous if they are the product of speciation and (3) paralogous if the product of gene duplication (for discussion of terminology, see Hall 1994).

Orthologous genes can be associated with homologous structures

One might expect some evolutionary stability in the genetic basis for particular aspects of morphology, and indeed many cases are known where orthologous genes are expressed in homologous structures in different taxa (Figs 2A & 3A). For instance, *engrailed* is expressed in homologous neurons within the central nervous systems of crustaceans and insects (Patel et al 1989), and *orthodenticle* is expressed in epithelial cells of podia in sea urchins and brittle-stars (Lowe & Wray 1997). In both cases, the taxa being compared diverged about half a billion years ago, suggesting that the association between homologous regulatory genes and homologous structures can persist for long periods. It is worth bearing in mind, however, that a clear evolutionary association between a homologous gene and a homologous structure has been documented primarily within phyla, and that there are relatively few unambiguous cases of such conservation among phyla. This situation may reflect the fact that most interphylum comparisons to date involve arthropods and chordates, two distantly related phyla that share relatively few

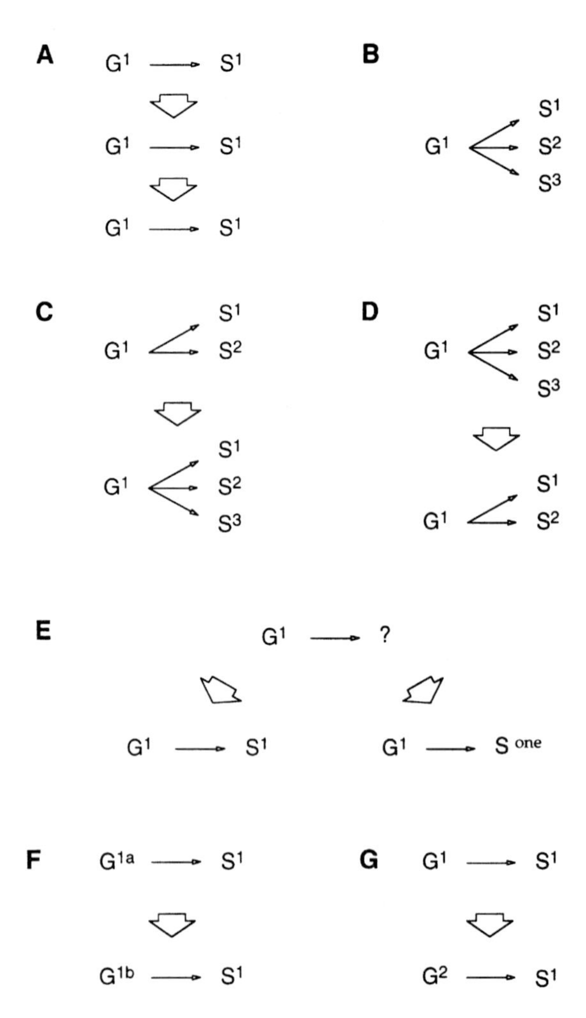

FIG. 2. Associations and dissociations between genes and structures. Associations between genes (G) and the morphological structures (S) whose development they regulate are diagrammed; development is indicated by thin arrows, evolution by thick arrows. (A) The association between gene and structure is often evolutionarily conserved, particularly over relatively short phylogenetic distances. (B) Most regulatory genes are required for the development of several distinct structures (S^1– S^3) in any given species, and these structures are often not homologous. (C, D) Regulatory genes can gain and lose roles in the development of particular structures during the course of evolution. (E) In some cases, homologous regulatory genes are involved in the development of analogous structures (S^1 and S^{one}). In such cases, it is not always clear what the gene was doing in the common ancestor. (F) Sometimes paralogous genes (G^{1a} and G^{1b}) are involved in the development of homologous structures, while their orthologues are not. (G) In more extreme cases, homologous genes are not required for the development of homologous structures, and non-homologous genes (G^1 and G^2) have apparently taken over the role. Probable examples of each case are discussed in the text.

specific homologous features. As data become available for more closely related phyla that share a greater number of derived morphological features, clear examples of conserved evolutionary associations between regulatory genes and structures that span phyla may accumulate.

Orthologous genes can be associated with non-homologous structures

The conservation of evolutionary association just discussed focuses on a single aspect of the total expression pattern of a regulatory gene. As reviewed earlier, however, most regulatory genes have several phases of expression (Fig. 2B). Distinct phases of expression typically occur in non-homologous structures and can involve unrelated developmental processes. For instance, in *Drosophila engrailed* is expressed in segmental stripes in the ectoderm, in specific neurons of both the central and peripheral nervous systems, in the fat bodies, in the gut primordium and in a variety of other locations (Duboule 1994, Rogers & Kaufman 1996). This association between a regulatory gene and several non-homologous structures seems to be the rule rather than the exception (Duboule & Wilkins 1998). The evolutionary implications of this one-to-many mapping (Fig. 2B) have received little attention. One clear message, however, is that the mere presence of an expression domain of a regulatory gene in two species is not a sure indicator that the structures where expression is occurring are homologous.

Orthologous genes can gain or lose an association with homologous structures

In some cases orthologous genes have an inconsistent evolutionary association with homologous structures (Fig. 2C,D; de Beer 1971, Roth 1988). These situations imply at least one evolutionary gain or loss of an association between that phase of expression and the structure in question (Fig. 3B,C). By mapping the presence and absence of the expression domain and structure onto a phylogeny, it is possible to identify independent gains and losses of roles and structures. For instance (Fig. 4), a segmentation role for *even-skipped* is present in some, but not all, insects; it probably evolved after segments themselves, and was subsequently lost at least once (Patel et al 1992, Grbic & Strand 1998). Many other likely gains and losses of developmental roles are implied by clear differences in expression domains among taxa (e.g. Patel et al 1989, Falciani et al 1996, Holland et al 1996, Savage & Shankland 1996, Lowe & Wray 1997). Recruitment and loss of developmental roles may be more common than generally appreciated, particularly when comparing taxa that diverged hundreds of millions of years ago. Although evolutionary dissociations between homologous genes and structures can clearly arise within phyla, they seem more common in comparisons among phyla. A general implication is that the expression of

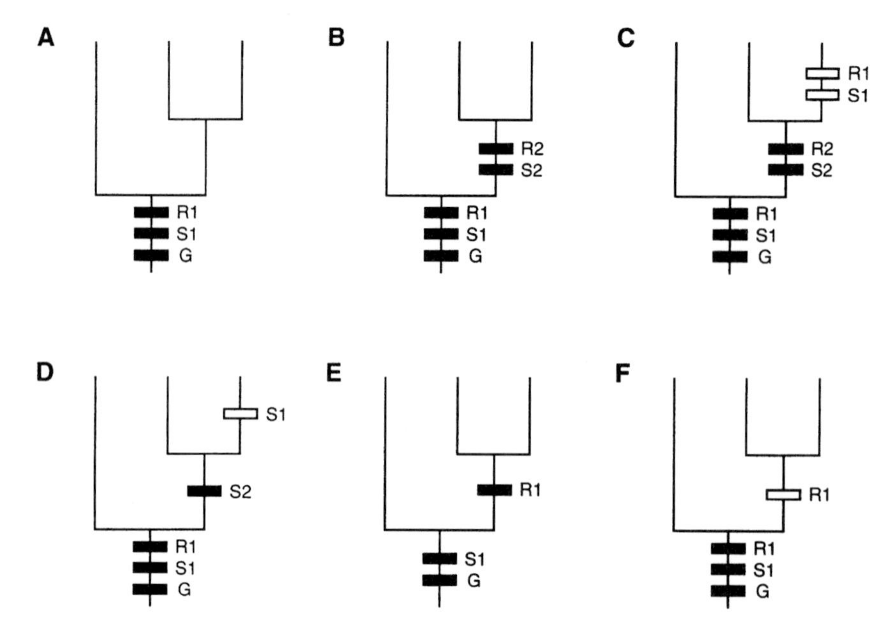

FIG. 3. Evolutionary scenarios for associations and dissociations between genes and structures. Mapping comparative data for genes (G), their developmental roles (R) and the structures in which they are expressed (S) onto a phylogeny reveals several distinct evolutionary patterns. Closed bars indicate the origin of a feature, and open bars indicate a loss. (A) Conservation of an ancient developmental role and structure. (B) Acquisition of an additional developmental role, a process known as recruitment. (C) Loss of a developmental role. (D) Recruitment of an existing developmental role to build a new structure. (E) Recruitment of a developmental role after the structure in which it is expressed has evolved. (F) Loss of a developmental role without loss of the structure in which it is expressed. Likely examples of all six evolutionary processes are known, and a few of these are discussed in the text.

regulatory genes will work best as indicators of homology over relatively short phylogenetic distances. Another implication is that the absence of an expression domain of a regulatory gene can provide only weak evidence that two structures are not homologous.

Orthologous genes can be associated with analogous structures

A surprising finding that has recently emerged is an association between homologous genes and what are superficially similar, but non-homologous, structures (Fig. 2E). Orthologues of *distalless* are expressed in the termini of developing limbs in arthropods and chordates, and in the termini of various outgrowths in other phyla (Panganiban et al 1997). At least some of these structures are clearly not homologous, suggesting that *distalless* was recruited into the development of body wall protrusions on several separate occasions. It

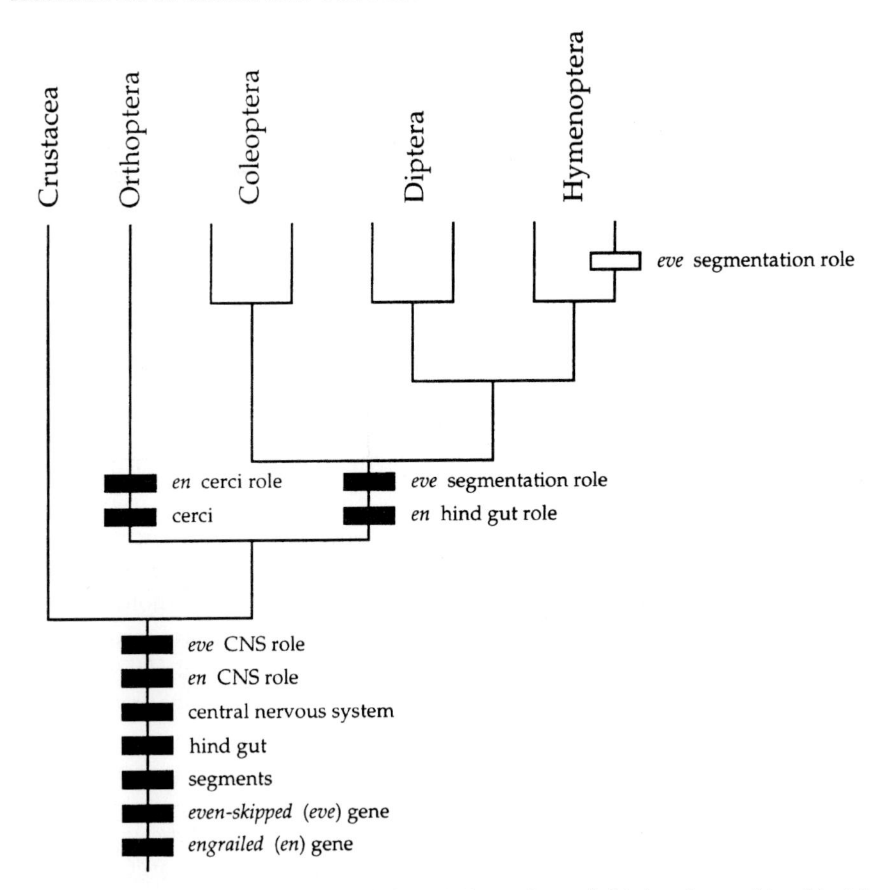

FIG. 4. Evolutionary association and dissociations of *engrailed* (*en*) and *even-skipped* (*eve*) in arthropods. A relatively extensive taxonomic sampling of expression data for *engrailed* and *even-skipped* exists for arthropods (Patel et al 1989, 1992, Rogers & Kaufman 1996, Grbic & Strand 1998, and references therein). A hypothesis for the evolution of these genes, their developmental roles and the structures in which they are expressed is shown, based on this comparative data. Several differences between taxa have been omitted for simplicity. Both genes are clearly older than the Arthropoda, based on the presence of orthologues in other phyla (Duboule 1994). Some roles for these genes probably date back to the early evolutionary history of the phylum, while others evolved during its diversification. Note that some roles evolved after the structures in which they are expressed, while at least one likely loss of an important developmental role has occurred. Closed bars indicate the origin of a feature, and open bars indicate a loss.

seems highly unlikely that this happened entirely by chance. Instead, *distalless* may be part of a proximodistal patterning system that was recruited as a unit during the evolutionary origin of protrusions (Fig. 3D). It makes sense that the evolution of novel structures might sometimes involve recruiting existing machinery for generic developmental tasks, such as patterning or cell signalling, rather than inventing them from scratch. Although it is difficult to assess how common this process is, other cases are known. For instance, nested domains of *Hox* gene expression specify position along the anteroposterior axis both in the brain and somites of vertebrates (Duboule 1994), which are clearly not homologous structures. An obvious implication is that genes involved in generic tasks may be positively misleading as indices of morphological homology.

Paralogous genes can be associated with homologous structures

Another interesting dissociation between gene and phenotype involves the expression of paralogous, rather than orthologous, genes in homologous structures (Fig. 2F). For example, *Wnt3a* is required for signalling by the apical ectodermal ridge during limb development in chickens but not in mice, where *Wnt10b* seems to fulfill this role (Greco et al 1996, Wang & Shackleford 1996, Kengaku et al 1998). It is not known how this evolutionary 'swapping' of paralogues occurs, but redundant roles for paralogues in an ancestor is an obvious possibility. Whatever its historical basis, this phenomenon seems likely to be less common than conserved association between orthologues and homologous structures. Thus, although it remains important to keep track of orthology when comparing expression domains (Abouheif et al 1997, Holland 1999, this volume), the possibility of paralogue swaps cannot be ignored.

Non-homologous genes can be associated with homologous structures

The most extreme dissociation between genotype and phenotype would be a replacement of the genetic basis for a particular phenotypic feature by non-homologous genes (Fig. 2G). This might seem unlikely on first principles. Furthermore, techniques for comparisons at the molecular level rely on sequence similarities and are therefore strongly biased against detecting such replacements. Nonetheless, several likely cases have been identified. One involves the sex determination and dosage compensation systems of nematodes, arthropods and chordates, which have few genes in common (Wilkins 1993). If, as generally assumed, these phyla evolved from an ancestor that had males and females, then extensive replacements of the genes involved in sex determination and dosage compensation must have occurred. The alternative explanation is that sex evolved independently in the lineages leading to these phyla. Even if this latter interpretation is correct, comparisons among closely related dipteran insects reveal dramatic changes: the 'master regulatory gene' in the *Drosophila* sex determination pathway is *sex lethal* (Wilkins 1993), but the homologous gene does not have this function in two other dipterans (Meise et al 1998, Saccone et al 1998). Because of the inherent technical difficulties involved in detecting such gene replacements, it impossible to know how commonly they occur. Knowing that gene replacements can occur, however, reinforces the importance of not relying on negative evidence when using gene expression to test hypotheses of homology at the morphological level.

Using gene expression to test hypotheses of morphological homology

There is currently considerable interest in using developmental regulatory genes to provide evidence for homology of structures. Using genes as a 'Rosetta Stone' (Slack 1984) to decipher the complex history of morphological change is an appealing idea. It is made all the more so by some successes using this approach. Notable among these are detailed analyses of regulatory gene expression in cephalochordates and urochordates, which have clarified homologies to the central nervous system of vertebrates (e.g. Holland et al 1992, Wada et al 1998). When combined with a careful consideration of morphological data and placed into a phylogenetic framework, expression data from regulatory genes can prove immensely illuminating (Shubin et al 1997).

It is clear, however, that the expression domains of homologous regulatory genes are not infallible guides to homology at the level of morphology (de Beer 1971, Roth 1988, Dickinson 1995, Bolker & Raff 1996, Abouheif 1997, Abouheif et al 1997). This was apparent as soon as data became available indicating that regulatory genes often have several distinct developmental roles in non-homologous structures within a single species. It has taken longer to learn that there are several different ways in which genes and morphology can become evolutionarily dissociated (Fig. 2C–F, and discussed above). Some of these dissociations can only be detected by comparing data from three or more taxa, which remains unusual in the literature. None the less, the data available already contain many examples of these various evolutionary dissociations, suggesting that they are not particularly rare events.

Dissociations between homologous genes and homologous structures do not invalidate the use of expression data to test hypotheses of morphological homology, but they do highlight some potential pitfalls. The fact that many regulatory genes have several developmental roles means that one must be careful to compare only the homologous role when testing hypotheses of morphological homology. To do otherwise leads to nonsensical inferences: the expression of *distalless* in the brain of amphioxus (Holland et al 1996) and in the eyespots of butterfly wings (Carroll et al 1994) does not imply that these are homologous structures. It may be relatively simple to identify homologous roles where morphological homologies are already clear, but this will often be difficult where they are not. Unfortunately, the latter cases may include those of greatest interest. The fact that developmental roles can be gained and lost introduces several complications. It means that one cannot simply assume that orthologous genes have the same set of developmental roles in two species. This will probably be true if the species are closely related, but may well be false if they are not. Again, however, comparisons of more distantly related taxa are more likely to be the most interesting cases. The existence of gains and losses of developmental roles also means that one must be exceedingly cautious about interpreting negative evidence. A regulatory gene that is expressed in a particular structure in one species may not be expressed in the genuinely homologous structure in another species. This could arise either because the developmental role evolved after the structure (Fig. 3E) or because the role was lost while retaining the structure (Fig. 3F). Perhaps most problematically, overt similarities in expression domains do not always reflect a common evolutionary origin, and may instead arise through convergence. Homologous genes can be independently recruited into the development of non-homologous, but superficially similar, structures. Such genes are probably the worst possible molecular tests of morphological homology, as they may be preferentially recruited into similar developmental roles on independent occasions. If this process is at all common, it will be important to formulate independent criteria for distinguishing between convergently similar expression domains and those that share a common evolutionary origin. A further difficulty is that portions of gene networks may be recruited as a unit in such cases (Abouheif 1999, this volume), which means that examining additional genes in the same network may effectively add no new evidence when testing hypotheses of morphological homology.

These pitfalls are serious, but not debilitating. If used judiciously, gene expression domains provide an extremely valuable source of evidence with which to test hypotheses of homology. The most meaningful inferences will be those based on (1) genes that are demonstrably orthologous, (2) gene expression domains that are demonstrably homologous and (3) tied to a plausible history of morphological transformations (Abouheif et al 1997). In general, regulatory genes will be most reliable as indices of morphological homology over relatively short time-scales. More information is becoming available about the evolution of regulatory genes, their expression domains, their developmental roles and the conditions under which they can become evolutionarily dissociated from particular morphological features. As we learn more about these evolutionary processes, it should become possible to use expression data to test hypotheses of morphological homology with increasing reliability.

Acknowledgements

Ehab Abouheif, Chris Lowe and Margaret Pizer provided many helpful comments. Work in my laboratory is supported by the National Science Foundation, the A. P. Sloan Foundation and the Stowers Institute for Medical Research.

References

Abouheif E 1997 Developmental genetics and homology: a hierarchical approach. Trends Ecol Evol 12:405–408
Abouheif E 1999 Establishing criteria for regulatory gene networks: prospects and challenges. In: Homology. Wiley, Chichester (Novartis Found Symp 222) p 207–225
Abouheif E, Akam M, Dickinson WJ et al 1997 Homology and developmental genes. Trends Genet 13:432–433
Arnone MI, Davidson EH 1997 The hardwiring of development: organization and function of genomic regulatory systems. Development 124:1851–1864
Bang AG, Bailey AM, Posakony JW 1995 *Hairless* promotes stable commitment to the sensory organ precursor cell fate by negatively regulating the activity of the *Notch* signaling pathway. Dev Biol 172:479–494
Bharathan G, Janssen BJ, Kellogg EA, Sinha N 1997 Did homeodomain proteins duplicate before the origin of angiosperms, fungi, and metazoa? Proc Natl Acad Sci USA 94:13749–13753
Bolker JA, Raff RA 1996 Developmental genetics and traditional homology. BioEssays 18:489–494
Carroll SB, Gates J, Keys DN et al 1994 Pattern formation and eyespot determination in butterfly wings. Science 265:109–114
Davis CA, Holmyard DP, Millen KJ, Joyner AL 1991 Examining pattern formation in mouse, chicken and frog embryos with an EN-specific antiserum. Development 111:287–298
de Beer GR 1971 Homology: an unsolved problem. Oxford University Press, Oxford (Oxford Biol Readers 11)
De Robertis EM, Sasai, Y 1996 A common plan for dorsoventral patterning in Bilateria. Nature 380:37–40
Dickinson WJ 1995 Molecules and morphology: where's the homology? Trends Genet 11:119–121
Duboule D (ed) 1994 Guidebook to the homeobox genes. Oxford University Press, Oxford
Duboule D, Wilkins AS 1998 The evolution of 'bricolage'. Trends Genet 14:54–59
Falciani F, Hausdorf B, Schröder R et al 1996 Class 3 *Hox* genes in insects and the origin of *zen*. Proc Natl Acad Sci USA 93:8479–8484
Falconer DS 1989 Introduction to quantitative genetics, 3rd edn. Longman, London
Finkelstein R, Boncinelli E 1994 From fly head to mammalian forebrain: the story of *otd* and *Otx*. Trends Genet 10:310–315
Franks RR, Hough-Evans BR, Britten RJ, Davidson EH 1988 Spatially deranged though temporally correct expression of a *Strongylocentrotus purpuratus* actin gene fusion in transgenic embryos of a different sea urchin family. Genes Dev 2:2–12
Grbic M, Strand MR 1998 Shifts in the life history of parasitic wasps correlate with pronounced alterations in early development. Proc Natl Acad Sci USA 95:1097–1101
Greco TL, Takada S, Newhouse MM, McMahon JA, McMahon AP, Camper SA 1996 Analysis of the vestigial tail mutation demonstrates that *Wnt-3a* dosage regulates mouse axial development. Genes Dev 10:313–324
Hall BK (ed) 1994 Homology: the hierarchical basis of comparative biology. Academic Press, San Diego, CA
Holland ND, Panganiban G, Henyey EL, Holland LZ 1996 Sequence and developmental expression of *AmphiDll*, an amphioxus *Distal-less* gene transcribed in the ectoderm, epidermis and nervous system: insights into the evolution of the craniate forebrain and neural crest. Development 122:2911–2920
Holland PWH 1999 How gene duplication affects homology. In: Homology. Wiley, Chichester (Novartis Found Symp 222) p 226–242

Holland PWH, Holland LZ, Williams NA, Holland ND 1992 An amphioxus homeobox gene: sequence conservation, spatial expression during development and insights into vertebrate evolution. Development 116:653–661

Kengaku M, Capdevila J, Rodriguez-Esteban C et al 1998 Distinct WNT pathways regulating AER formation and dorsoventral polarity in the chick limb bud. Science 280:1274–1277

Kimmel CB 1996 Was Urbilateria segmented? Trends Genet 12:329–331

Klueg KM, Harkey MA, Raff RA 1997 Mechanisms of evolutionary changes in timing, spatial expression, and mRNA processing in the *msp130* gene in a direct-developing sea urchin, *Heliocidaris erythrogramma*. Dev Biol 182:121–133

Lowe CJ, Wray GA 1997 Radical alterations in the roles of homeobox genes during echinoderm evolution. Nature 389:718–721

Ludwig MZ, Patel NH, Kreitman M 1998 Functional analysis of *eve* stripe 2 enhancer evolution in *Drosophila*: rules governing conservation and change. Development 125:949–958

Mastick GS, McKay RM, Oligino TJ, Donovan K, López AJ 1995 Identification of target genes regulated by homeotic proteins in *Drosophila melanogaster* through genetic selection of *Ultrabithorax* protein binding sites in yeast. Genetics 139:349–363

Meise M, Hilfikerkleiner D, Dubendorfer A, Brunner C, Nothiger R, Bopp D 1998 *Sex-lethal*, the master sex-determining gene in *Drosophila*, is not sex-specifically regulated in *Musca domestica*. Development 125:1487–1494

Moskowitz IP, Rothman JH 1996 *lin-12* and *glp-1* are required zygotically for early embryonic cellular interactions and are regulated by maternal GLP-1 signaling in *Caenorhabditis elegans*. Development 122:4105–4117

Panganiban G, Irvine SM, Lowe C et al 1997 The origin and evolution of animal appendages. Proc Natl Acad Sci USA 94:5162–5166

Patel NH, Martin-Blanco E, Coleman KG et al 1989 Expression of *engrailed* proteins in arthropods, annelids and chordates. Cell 58:955–968

Patel NH, Ball EE, Goodman CS 1992 Changing role of *even-skipped* during the evolution of insect pattern formation. Nature 357:339–342

Rogers BT, Kaufman TC 1996 Structure of the insect head as revealed by the EN protein pattern in developing embryos. Development 122:3419–3432

Roth VL 1988 The biological basis of homology. In: Humphries CJ (ed) Ontogeny and systematics. Columbia University Press, New York, p 1–26

Saccone G, Peluso I, Artiaco D, Giordano E, Bopp D, Polito LC 1998 The *Ceratitis capitata* homologue of the *Drosophila* sex-determining gene *sex-lethal* is structurally conserved, but not sex-specifically regulated. Development 125:1495–1500

Savage RM, Shankland M 1996 Identification and characterization of a *hunchback* orthologue, *Lzf2*, and its expression during leech embryogenesis. Dev Biol 175:205–217

Schumacher A, Magnuson T 1997 Murine *Polycomb-* and *trithorax*-group genes regulate homeotic pathways and beyond. Trends Genet 13:167–170

Shubin N, Tabin C, Carroll S 1997 Fossils, genes and the evolution of animal limbs. Nature 388:639–648

Slack JMW 1984 A Rosetta stone for pattern formation in animals? Nature 310:364–365

Slack JMW, Holland PWH, Graham CF 1993 The zootype and the phylotypic stage. Nature 361:490–492

Wada H, Saiga H, Satoh N, Holland PWH 1998 Tripartite organization of the ancestral chordate brain and the antiquity of placodes: insights from ascidian *Pax-2/5/8*, *Hox*, and *Otx* genes. Development 125:1113–1122

Wang J, Shackleford GM 1996 Murine *Wnt10a* and *Wnt10b*: cloning and expression in developing limbs, face and skin of embryos and in adults. Oncogene 13:1537–1544

Wilkins AS 1993 Genetic analysis of animal development, 2nd edn. Wiley-Liss, New York

Wright S 1980 Genic and organismic selection. Evolution 34:825–843

Zuckerkandl E 1994 Molecular pathways to parallel evolution: I. gene nexuses and their morphological correlates. J Mol Evol 39:661–678

DISCUSSION

Wagner: Multicellular organs are not made by genes but by cells, and genes regulate cell behaviour. If you think in terms of such a cellular context, much of what you said becomes easier to digest. For example, mesenchyme–epithelium interactions are involved in localizing rudiments for organs in chordates, and if we know about what generic cellular processes are involved in rudiment formation then we will have a theoretical framework in which the relationship between genes and phenotypes becomes more understandable.

Wray: It's a matter of perspective. I like your term 'generic' because that's exactly what's going on. There are only a few genes that encode signalling molecules in the genome, particularly relative to the number of signalling events during development. Clearly, the same process in being used over and over again.

Akam: I don't think organisms care much about genes. They don't care which gene is encoding a particular function; what matters is that the cell carries out the function. Whether that function comes from one gene or another may be quite labile, particularly in vertebrates, where there is extensive, if partial, redundancy of proteins encoded by duplicated genes. Even if you are aware of which genes are orthologous and which genes are paralogous, you may find that conserved roles are switching between paralogous genes.

Wray: I was afraid to state things quite that boldly, but I agree.

Rudolf Raff: There are data that suggest this is occurring in multigene families. A good example involving *Wnt* genes will be discussed by Ehab Abouheif in his presentation (Abouheif 1999, this volume)

Tautz: A few years ago people had similar arguments, i.e. they were stating that genes are useful for tracing homologies, and we now know many genes that are useful for doing this. You presented a strong argument for analysing genes because you argued that they don't have to have the same roles and the same structures, but in those cases where they do, they are extremely useful. For example, *even-skipped* has erratic expression patterns in insects, and should not be used as a marker for tracing segmental origins, but in contrast *engrailed* is expressed at the second boundary in all arthropods tested so far (Patel et al 1989, Holland et al 1997), so there's a strong argument that if we find a similar pattern of expression in chordates then it must mean something.

Wray: You have adopted the position of a morphologist, in the sense that a morphologist would say 'there are some morphological structures that I know are homoplasic, therefore I will not use them to build my phylogenies, and there are others that I can trust, so I will use those instead'. The history of morphology does inform us about the history of the roles of these genes. The fact that these genes have complex histories is interesting to study because they are real phenomena.

Rudolf Raff: There seems to be a dichotomy of interest here. We want to trace evolutionary histories, in which case we need to analyse genes which have expression patterns that enable us to do this. However, we also want to find processes that generate novelties, and in order for this to happen we have to break away from strictly homologous events into the fascinating processes of homology. You have broken down many of the co-option and dissociation events into more specifically defined categories, which will be helpful in taking these analyses further.

Striedter: I was intrigued by your statement that you rarely see convergence discussed in the context of gene expression, and that you expected to see convergence more often in comparisons of distantly related groups. What strikes me about parallelism is that it generates things that look almost exactly identical, so would you expect parallel gene expression patterns to be more common in closely related organisms?

Wray: I don't know what to predict. So far, the use of morphology to predict the kinds of phenomena we might expect to find has been a pretty good guide, so it is possible that parallel gene expression is more common in closely related groups, but it may be difficult to detect.

Greene: If I had two cages of wild-type mice, and I came in one morning to find a white mouse in each cage, would that white coat colour be homologous or homoplasic?

Akam: A mouse geneticist knows how many different genes can mutate to give a white coat. In the case of *Drosophila*, if you find a white-eyed fly in two cages, those mutations are homologous in the sense that the mutations will have occurred in the same gene. On the other hand, many genes can mutate to give rise to a brown eye colour. We know that for most complex traits exactly the same phenotype can be generated by completely separate mutations, so we would predict that for most traits, even when you see a similar phenotypic change, the genetic basis for that change is unlikely to be the same.

Wilkins: Julian Huxley wondered about this more than 60 years ago. He said there are white-eyed *Drosophila virilis* mutants and white-eyed *Drosophila melanogaster* mutants, and we know that these strains didn't come from the same parents; however, can we say that the 'white eye' information is homologous? He concluded that they can't be homologous because genealogically the parents are not the same species. This is no longer a conceptual problem because one can be certain he was looking at mutations in the same gene in the two different species.

Greene: What if I had two closely related species in the two cages?

Wilkins: They will probably have mutations in the same gene. It is the same genetic information, but the phylogenetic use of homology falls down.

Rudolf Raff: It doesn't fall down at all. The definition of homology requires 'shared and derived' characteristics and therefore the two white mice are not homologous.

Akam: I don't agree.

Meyer: I would like to bring up the issue of homologous roles. I was surprised that in that context you used the words 'functional homology' because we should try to avoid this idea, since 'function' is not a defining characteristic of homologous traits.

Wray: Function is not a criterion for recognizing homology, but that doesn't mean functions cannot be conserved and therefore homologous.

Müller: Wouldn't it be better to homologize between functions rather than to deduce morphological homology from functional homology?

Wray: Absolutely.

Müller: Diethard Tautz pointed out earlier that if *engrailed* is found to be expressed in chordates, then this must mean something. But does this mean that the morphological structures which result from the segmentation process are homologous, or isn't rather the process homologous? This is an important distinction we have to make if we do not want to become totally confused about the kind of homology we are speaking of.

Tautz: No, it does not mean this at that point. It just means that we should look in more detail.

Akam: But even if the entire set of interactions have been conserved we are still left with two possibilities. One is that the structures are homologous; and the other is that you have had recruitment of a network of genes. We don't have the data at the moment to say which of those is the more likely.

Tautz: This is relevant to Günter Wagner's idea of having entities that are homologues, and these entities could be the reflection of genes, which is what we can trace.

Abouheif: I would still argue that there's a need to keep the levels separate. For example, a gene such as *distalless*, which is expressed at the distal tip of the appendages in arthropods, echinoderms and chordates, may be homologous as genes and as homologous processes, but they are doing so in non-homologous structures.

Tautz: I would like to say a few words about *distalless*. It was cloned in *Drosophila* because there was a mutant for it that was obtained by chance. However, there was no systematic screen for appendage mutants at that time. And it is also clear that this is difficult to perform because the appendages are first made as precursors (imaginal discs). Any specific phenotype in them becomes only apparent after metamorphosis and many mutants do not make it to this stage. Thus, almost all genes known to be involved in appendage formation in *Drosophila* were cloned because of other reasons and their role in the appendages is secondary. *distalless* has now been found to be a highly conserved gene and is detected in many forms of outgrowth structures (Panganiban et al 1997). But how many genes are really involved in forming the appendages in *Drosophila*? How many will eventually show up to be conserved? Therefore, one should be careful about any conclusions from seeing *distalless* alone to be expressed in such outgrowth structures. Only a full picture with the knowledge about the other genes involved in this process will eventually allow us to call such structures homologous.

Rudolf Raff: We have to be extremely careful about the level of homology because we have homologous genes and we are going to have homologous gene networks. There then has to be recognition at the next level, i.e. the structural level. It may well be that a homologous gene network plays a role in a non-homologous structure. If we are not precise about these levels we will confuse ourselves.

Galis: But how much of the gene networks are the same? In mice the *Hox* genes function everywhere, but the different functions are regulated independently and are interacting with different genes. Therefore, it's not really that the same genetic network operates in each case, but there are similarities.

Akam: We need to distinguish between protein networks — i.e. proteins that work together physically and are going to be recruited together because the proteins do not re-evolve — and gene networks, i.e. genes encoding proteins that interact with a DNA site that regulates another gene. I am convinced that the latter will be more labile. Protein–DNA interactions do re-evolve.

Wagner: Perhaps we become confused because analysis at the gene level is not the correct level of analysis, but rather it is protein functional domains or the domains of regulation that are the functional units we need to look at.

Akam: We have also not talked about cell behaviour, and to jump straight from a gene network to a morphological structure I think is missing the analysis at the level of the cell.

References

Abouheif E 1999 Establishing homology criteria for regulatory gene networks: prospects and challenges. In: Homology. Wiley, Chichester (Novartis Found Symp 222) p 207–225

Holland LZ, Kene M, Williams NA, Holland ND 1997 Sequence and embryonic expression of the amphioxus *engrailed* gene (AmphiEn): the metameric pattern of transcription resembles that of its segment-polarity homolog in *Drosophila*. Development 124:1723–1732

Panganiban G, Irvine SM, Lowe C et al 1997 The origin and evolution of animal appendages. Proc Natl Acad Sci USA 94:5162–5166

Patel NH, Martin-Blanco E, Coleman KG et al 1989 Expression of *engrailed* proteins in arthropods, annelids and chordates. Cell 58:955–968

Establishing homology criteria for regulatory gene networks: prospects and challenges

Ehab Abouheif

Department of Ecology and Evolution, State University of New York at Stony Brook, Stony Brook, NY 11794-5245, USA

Abstract. One of the most remarkable discoveries to emerge from the field of developmental genetics is the observation that many regulatory genes and segments of their interactive networks (pathways) appear to have been conserved in several metazoan phyla. Determining whether these conserved regulatory networks are homologous, i.e. derived from an equivalent network in the most recent common ancestor, is critical to understanding comparisons between model system studies, and the evolution of metazoan body plans. To this end, I outline some of the evolutionary properties of regulatory networks, and propose both similarity and phylogenetic criteria that can be used to test the hypothesis that two regulatory networks are homologous. Furthermore, I propose that genetic networks can be treated as a distinct level of biological organization, and can be analysed together with other hierarchical levels, such as genes, embryonic origins and morphological structures, in a comparative framework. Examples from the literature, particularly the genetic regulatory networks involved in patterning arthropod and vertebrate limbs, are examined using the proposed criteria and hierarchical approach.

1999 Homology. Wiley, Chichester (Novartis Foundation Symposium 222) p 207–225

The last two decades of research in the field of developmental genetics has revealed that many of the genes which control embryonic development are remarkably conserved across a broad range of metazoan phyla (De Robertis 1994). Recent comparisons between arthropods, nematodes and chordates suggest that this conservation not only applies to developmental genes, but also to some of the regulatory interactions between them (Gerhart & Kirschner 1997). The discovery of evolutionarily conserved developmental genes and their interactive regulatory networks (pathways) has helped to revitalize the study of the connection between development and evolution, and has facilitated comparisons between distantly related model organisms.

Comparisons between conserved genes and networks must, however, be placed within an evolutionary framework (Abouheif 1997, Abouheif et al 1997). The concept of homology, herein defined as the derivation of an attribute from the most recent common ancestor of two organisms (Mayr 1982), provides such a framework and guides the observations, interpretations and conclusions drawn from any comparison made between two regulatory gene networks. Although there exist criteria for identifying homology among genes, as well as embryonic and morphological structures (Hall 1994), there has been little or no discussion of criteria for establishing the homology among regulatory gene networks. Genetic networks play an important role in the evolution of the genes which compose them, as well as the embryonic and morphological structures to which they give rise (Zuckerkandl 1994). Furthermore, genetic networks can be recognized as a distinct level of biological organization separate from genes, embryonic origins and morphology. Thus, the objective of this chapter is to outline some evolutionary properties of regulatory gene networks, and to establish both similarity and phylogenetic criteria for recognizing whether two genetic networks are homologous.

Searching for and discovering homologies among developmental genes and their interactive networks marks a new era of investigation. Although there currently exist few comparative data on regulatory gene networks, recent advances in molecular and developmental biology are allowing rapid progress in this area. This next century promises to be as exciting as the one in which classical morphologists, such as Etienne Geoffroy Saint-Hilaire, George Cuvier and Richard Owen, first began their search for the homology of anatomical structures across the animal kingdom (Mayr 1982).

Evolutionary properties of genetic networks

Any regulatory gene network is composed of a set of interacting genes (Fig.1; Arnone & Davidson 1997). Therefore, any consideration of whether two networks are homologous must be explicitly based on both the component genes and their regulatory interactions. In contrast to criteria for recognizing homology at other levels of biological organization, such as genes and morphology (Patterson 1988), understanding the interactions and linkages between the component genes in a network is of crucial importance. Considering the homology of both genes and their interactions is a unique property of regulatory networks.

Therefore, when evaluating comparisons between regulatory gene networks, one must distinguish between three alternative hypotheses: homology, convergence and partial homology (Fig. 2; Zuckerkandl 1994). For two regulatory networks to be homologous, all the genes and their interactions must be derived from a network in the most recent common ancestor (Fig. 2A). In

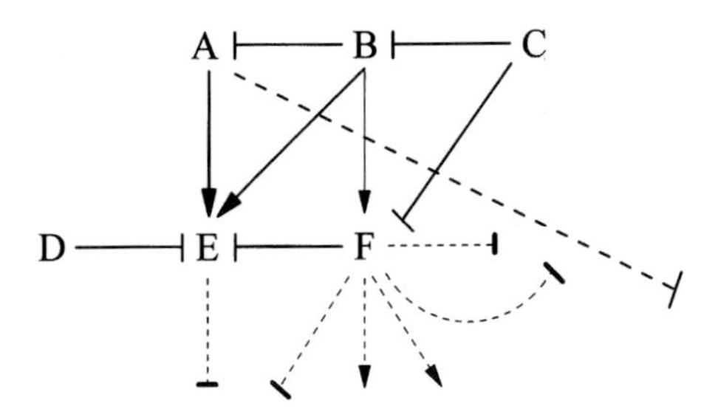

FIG. 1. A hypothetical regulatory gene network composed of six genes. The solid black lines indicate interactions between the genes A, B, C, D, E and F, whereas the dotted lines indicate interactions with genes in another region of the hypothetical regulatory network. Arrow heads indicate that the gene is activating another gene in the network, whereas the short solid lines indicate a repressive interaction.

contrast, for networks to be convergent all the genes and their interactions must have been recruited into a regulatory network after the divergence from a most recent common ancestor (Fig. 2B). Finally, for networks to be partly homologous, some genes and their interactions must be derived from the most recent common ancestor, whereas others must have been recruited into the pathway since the divergence of the species being compared (Fig. 2C).

At least three kinds of regulatory linkages can be involved in the evolution of genetic networks: obligate, facultative and novel linkages. Some members of a genetic network may be obligately linked. For example, tissues or cells expressing the signalling molecule *hedgehog* can, as far as known, only be perceived by the receptor molecule *patched* (Goodrich et al 1996). Currently, the only clear examples of obligate linkage are between genes that transmit signals within and between cells, particularly ligands and their receptors. The more obligate linkages there are in a regulatory gene network, the more likely it is to be conserved.

A particular gene in a regulatory network is facultatively linked if it can be functionally substituted by another gene through evolutionary time (Shubin et al 1997). Although a functional substitution event can occur between two unrelated genes, it may be more likely to occur between two members of a multigene family (i.e. between paralogues—gene copies produced through gene duplication; Xue & Noll 1996). For example, some members of the vertebrate Wnt gene family play important roles in limb morphogenesis. In the chick, the gene *Wnt3a* is essential for the formation of the apical ectodermal ridge (AER), a specialized ectodermal structure running along the distal margin of the developing limb bud (Kengaku et al 1998). The mouse orthologue of *Wnt3a* (i.e. a gene copy derived from the ancestral chick and mouse *Wnt3a* gene) is not expressed in the mouse AER, and the disruption of this gene does not affect AER formation (Greco et al 1996).

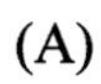

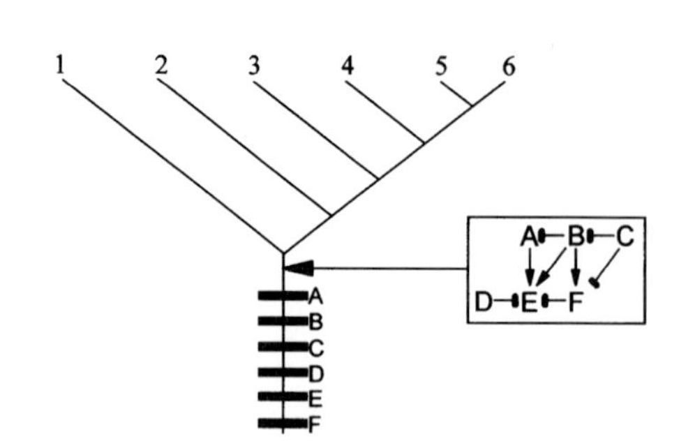
1
2
3
4
5
6
A B C
D E F
A
B
C
D
E
F

(B)

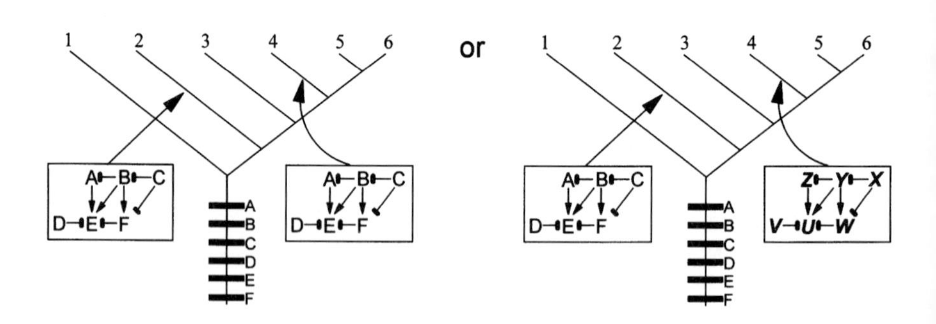
1
2
3
4
5
6
or
A B C
D E F
A B C
D E F
A
B
C
D
E
F
Z Y X
V U W

(C)

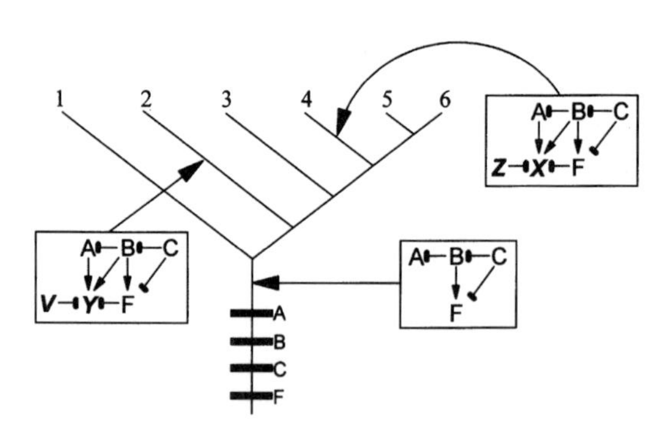
1
2
3
4
5
6
A B C
Z X F
A B C
V Y F
A B C
F
A
B
C
F

Interestingly, it is the mouse *Wnt10b* gene which has an expression domain similar to that of chick *Wnt3a* (Wang & Shackleford 1996). Depending on what the ancestral condition was, there may have been a functional substitution of *Wnt3a* by *Wnt10b* in the mouse, or a functional substitution of *Wnt10b* by *Wnt3a* in the chick. The existence of facultative linkages within regulatory gene networks, as in the above example, will tend to make networks partly homologous.

Finally, an inherent property of developmental regulatory genes is their ability to acquire novel regulatory linkages and developmental roles through evolutionary time (Gerhart & Kirschner 1997, Lowe & Wray 1997, Duboule & Wilkins 1998). This can increase the number of components of an existing regulatory network (Zuckerkandl 1997). For example, the transcription factor *distalless* has presumably acquired its multiple developmental roles — such as proximodistal axis formation of limbs in arthropods, echinoderms and chordates (Panganiban et al 1997), pattern-forming functions of the eyespots on butterfly wings (Carroll et al 1994) and branchial arch development in chordates (Akimenko et al 1994) — through the creation of novel regulatory linkages. Thus, the recruitment of novel genes into an existing regulatory gene network will also tend to make genetic networks partly homologous.

It is now well established that both morphological structures and genes can be recruited (co-opted) into new functions through evolutionary time (Gould & Vrba 1982, Lowe & Wray 1997). Thus, it is appropriate to ask whether regulatory gene networks, as a whole, can also be recruited to function in different developmental contexts and unrelated morphological structures. Current evidence suggests that this is indeed possible. For example, the Ras-RTK (receptor tyrosine kinase) pathway is a ubiquitous signal transduction network that has been highly conserved in nematodes, arthropods and chordates (Gilbert 1997). It is composed of approximately six core proteins that transmit signals between the cell surface and genes within the nucleus. This regulatory network appears to have been recruited to function in strikingly different morphological structures, such as mammalian skin, the nematode vulva and the *Drosophila* eye (Gilbert 1997). This implies that homology among networks cannot always be taken as strong evidence to support homology among

---

FIG. 2. Homology, convergence and partial homology among regulatory gene networks depicted on hypothetical six taxon trees, where the presence of the genes are indicated by black boxes. The interactions between these genes are shown in the little boxes. (A) A scenario of homology among the genes and their functional interactions. (B) A scenario of convergence by either recruitment of old genes into new regulatory interactions, or recruitment of new genes into an ancient set regulatory interactions. (C) Partial homology. Some elements of the regulatory network are homologous, whereas others have been recruited into the network since the divergence of the two species being compared.

embryonic origins and morphological structures. Although the Ras-RTK signal transduction network is homologous in fruit flies and nematodes, this clearly cannot be taken as evidence to support the homology between the nematode vulva and the *Drosophila* eye. Thus, recruitment is an evolutionary phenomenon that can be extended to regulatory gene networks, and further supports the notion that networks can be considered a distinct level of biological organization.

Establishing similarity and phylogenetic criteria for regulatory gene networks

The first step in testing whether two regulatory gene networks are homologous is to use a set of established similarity criteria to *propose* a hypothesis of homology, partial homology or convergence (Shubin 1994). In the following section, I suggest three similarity criteria for regulatory gene networks. Once a hypothesis has been proposed on the basis of these similarity criteria, the hypothesis must then be *tested* by mapping the individual elements of the regulatory gene network on a phylogenetic tree using the principles of parsimony (Patterson 1982, 1988). This phylogenetic criterion, discussed in more detail below, is used to test whether some or all the elements in a regulatory gene network were present in the common ancestor of the taxonomic groups being compared (homology), or whether some or all of the elements in the network have independent evolutionary origins (convergence).

Similarity criteria: proposing hypotheses of homology, partial homology and convergence

Defining regulatory gene networks. Arnone & Davidson (1997) define a genetic network as the regulatory regions of all the component genes and the interactions between them. Regulatory interactions between developmental genes are remarkably complex, and at present there exist only limited data on the operation of genetic networks (Yuh et al 1998). Therefore, it is important to be explicit about defining which subset of a genetic network one is interested in comparing when proposing a hypothesis of whether two regulatory gene networks are homologous, partly homologous or convergent.

For instance, the regulatory gene network shown in Fig. 1 is, in fact, a small subset of a much larger regulatory network (Fig. 3). This means that one is arbitrarily choosing or defining boundaries around a non-linear and continuous network. Because some regions of a regulatory network may be completely conserved, whereas others may be extensively modified over evolutionary time, it is important to attempt to define several different boundaries in order to see how sensitive the conclusions are to the boundaries chosen. Explicitly defining subsets of regulatory gene networks, although arbitrary, is currently the only

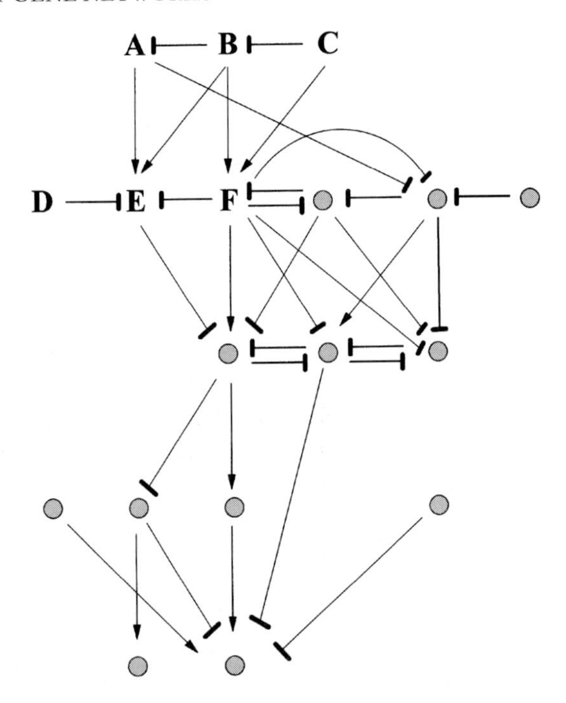

FIG. 3. The hypothetical gene network in Fig. 1 is shown with its connections to other genes to demonstrate the continuity and non-linearity among regulatory gene networks.

option available if we are to attempt to identify homologous elements and make meaningful comparisons across taxonomic groups.

Similarity of genes in a regulatory network. In order for two regulatory gene networks to be homologous, the genes within each network being compared must be orthologues of one another (gene copies produced through speciation; Abouheif et al 1997). Inferences of orthology should not be based on functional or sequence similarity, because unrelated genes can converge on similar genetic functions through evolutionary time, and measures of sequence similarity cannot distinguish between orthologues and paralogues (gene copies produced through gene duplication; Zardoya et al 1996). Determining the orthology between genes requires reconstructing the evolutionary history of the gene families being compared. Determining orthology in this way, however, becomes complicated when gene duplication events have occurred after the divergence of the species being compared (Holland 1999, this volume). For instance, there is only one copy of the *distalless* gene in *Drosophila*, but approximately seven copies of *distalless* in vertebrates. In such cases, one to one comparisons between genes in the regulatory networks of interest must be made with caution (Abouheif et al 1997).

Similarity of gene interactions. Similarity between (1) the biochemical function of the genes, (2) the developmental function of the genes and (3) the relative spatial position in which the genes are expressed, may indicate homology of the interactions among genes between two regulatory networks. For instance, the biochemical role of the *Drosophila hedgehog* gene is as a signalling molecule, and its developmental function in the developing wing imaginal disc is to specify anteroposterior fate by stimulating the secretion of Dpp protein (Brook et al 1996). This is accomplished by binding of Hedgehog protein to the patched receptor, alleviating the repression of Dpp by patched. These functional interactions are conserved in the chick (Shubin et al 1997), as the vertebrate gene *Sonic hedgehog*, is also a signalling molecule which stimulates the secretion of *BMP2*, a homologue of the *Drosophila dpp* gene. Furthermore, both *hedgehog* and *Sonic hedgehog* are expressed in the same relative spatial position, i.e. the posterior compartment in the developing imaginal disc and developing chick limb bud (Shubin et al 1997).

Phylogenetic criteria: testing hypotheses of homology, partial homology and convergence

Testing hypotheses of homology, partial homology and convergence proposed on the basis of the above similarity criteria requires use of the principles of parsimony to map the genes and their interactions onto a phylogenetic tree (for in depth discussion of how to apply this phylogenetic criterion see Kitching et al 1998; Fig. 2). For each species, all of the genes and interactions in the network should be analysed individually and defined as discrete taxonomic characters. When the characters (i.e. the genes and their interactions) are mapped onto a phylogenetic tree, they may support a hypotheses of homology—conservation dating back to their latest common ancestor (Fig. 2A). Alternatively, all or some of the characters may support a hypothesis of convergence (Fig. 2B) or partial homology (Fig. 2C).

The power of testing hypotheses of homology in this way depends upon the veracity of the phylogenetic tree used for the test, the number of taxa sampled and on the quality of the comparative data that are being mapped onto the tree. Furthermore, as knowledge on the operation of regulatory gene networks accumulates, it will be possible to incorporate this knowledge into the evolutionary assumptions made to map the characters (i.e. the genes and their interactions) onto the tree. For example, based on the limited amount of comparative data, it appears that pure convergence among regulatory networks may be improbable (Shubin et al 1997). The evolutionary gain of an entirely new regulatory gene network from its component genes appears less likely than the evolutionary loss of a network. This is due to the fact that only one upstream repressive interaction has to evolve for the regulatory network to be lost, whereas many new interactions must evolve to assemble an entirely new regulatory network. Thus, it is important to incorporate and account for these types of biological realities when applying the phylogenetic criterion to regulatory gene networks.

Hierarchical analyses of homology and regulatory gene networks

It is now reasonably well established that homology is a phenomenon that can be expressed independently at several distinct levels of biological organization (de Beer 1971, Roth 1988, Dickinson 1995), such as genes, gene functions, gene networks, as well as embryonic origins and morphological structures. Furthermore, homology at one level of biological organization does not necessitate homology at other levels (de Beer 1971, Dickinson 1995, Abouheif 1997, Abouheif et al 1997, Wray 1999, this volume). Because regulatory gene networks can be considered as a distinct level of biological organization, they can be analysed together with the other hierarchical levels using the principles of parsimony and the phylogenetic criterion described above (Lauder 1994, Abouheif 1997). This hierarchical approach can potentially reveal several interesting evolutionary scenarios. Ten, of many possible, scenarios and their implications are outlined in Table 1, which is presented in the hope that it will serve as a method of classifying and interpreting some of the more complex examples in the literature.

Comparing the regulatory network for patterning the dorsoventral axis of the wings of *Drosophila* and chick will demonstrate the utility of Table 1, and provide an example of the hierarchical approach. *Drosophila* and chick wings are convergent as morphological structures, and are derived from different embryological origins (Shubin et al 1997). The *Drosophila* wing is derived from imaginal disc cells in the larvae, whereas the chick wing is derived from a limb bud. Furthermore, the cell interactions that give rise to the *Drosophila* wing occur within the ectoderm, whereas the interactions that give rise to the chick wing are derived from both the ectoderm and mesoderm (Brook et al 1996).

Based on the similarity criteria listed above, I propose that the genetic regulatory network that patterns the *Drosophila* and chick wings may be partly homologous. Presently, there are insufficient comparative data to test this hypothesis using the phylogenetic criterion. Thus, alternative hypotheses, such as homology and convergence, should continue to be considered until enough comparative data are accumulated to distinguish between these possibilities. The dorsoventral network in *Drosophila* wings is shown in Fig. 4 (reviewed in Brook et al 1996). The selector gene *apterous* is expressed in and specifies the dorsal compartment of the imaginal wing discs (Fig. 4a). Both *apterous*- and *fringe*-expressing cells in the dorsal compartment then activate the gene *serrate* at the boundary of cells

TABLE 1 Ten evolutionary scenarios resulting from mapping different levels of biological organization, including gene networks, on a phylogenetic tree

Scenario \ *Biological level*	*Genes*	*Gene interactions*	*Gene network*	*Embryonic origin*	*Morphological structure*	*Phylogenetic representation*[a]	*Evolutionary implications*
1	H	H	H	H	H	1 2 3 4 5 6; m c gn gi g	Homology at all levels of biological organization.
2	H	H	H	C	C	1 2 3 4 5 6; m c; m c; gn gi g	Recruitment of homologous genes and network to function in convergent embryonic and morphological structures.
3	some H some C	H	PH	C	C	1 2 3 4 5 6; m c g; m c g; gn is PH; gi g	Recruitment of novel genes into an ancestral genetic network, and recruitment of this network to function in convergent embryonic and morphological structures.
4	H	some H some C	PH	C	C	1 2 3 4 5 6; m c gi; m c gi; gn is PH; gi g	Recruitment of novel regulatory interactions into an ancestral network, and recruitment of this network to function in convergent embryonic and morphological structures.
5	H	C	C	C	C	1 2 3 4 5 6; m c gn gi; m c gn gi; g	Ancient and homologous genes recruited to interact in a convergent network as well as convergent embryonic origins and morphological structures.

Scenario \ Biological level	Genes	Gene interactions	Gene network	Embryonic origin	Morphological structure	Phylogenetic representation[a]	Evolutionary implications
6	some H some C	H	PH	H	H	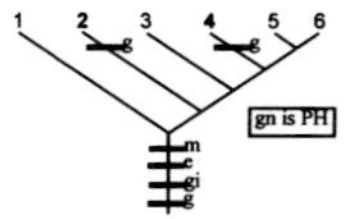 gn is PH	Recruitment of novel genes into an ancestral network that gives rise to homologous embryonic and morphological structures.
7	H	some H some C	PH	H	H	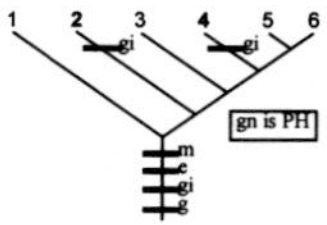 gn is PH	Recruitment of novel regulatory interactions into an ancestral network that gives rise to homologous embryonic and morphological structures.
8	C	C	C	H	H	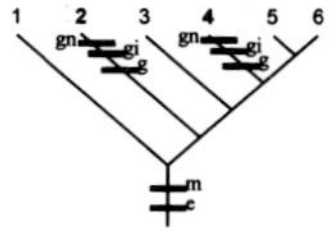	A convergent regulatory gene network giving rise to homologous embryonic and morphological structures.
9	C	C	C	C	H	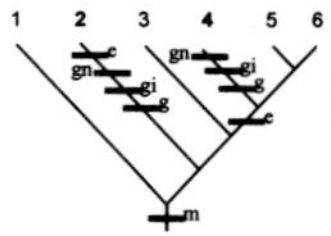	Homologous morphological structure maintained, in spite of other levels being convergent.
10	C	C	C	C	C	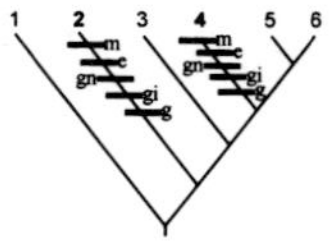	Convergence on all levels of biological organization.

[a]Different levels of biological organization are mapped on a hypothetical six-taxon tree. Each character is represented by a black box, and is labelled according to which level of biological organization it represents. C, convergent; e, embryonic origins; g, genes; gi, gene interactions; gn, gene network; H, homologous; m, morphological structure; PH, partly homologous.

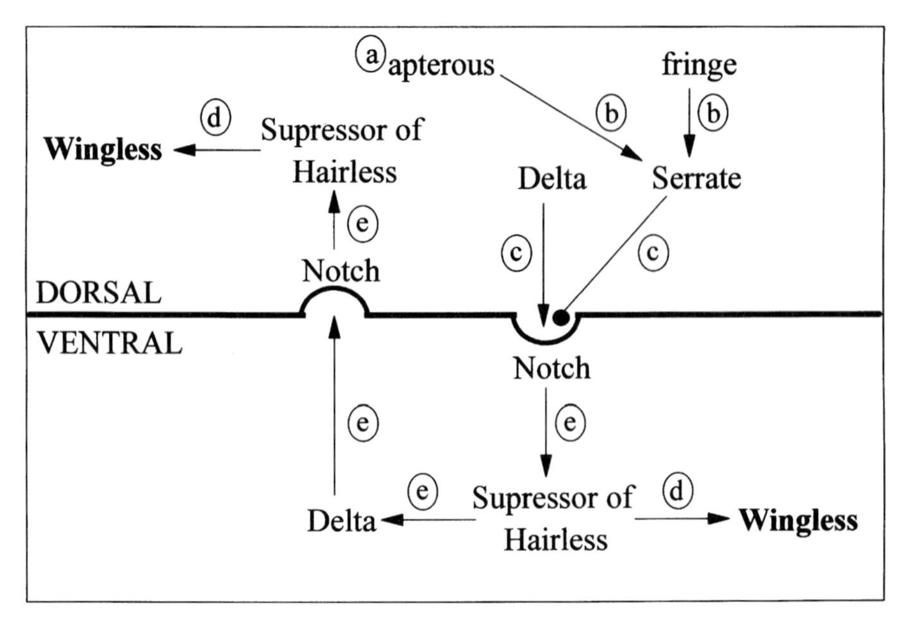

FIG. 4. The regulatory gene network involved in setting up the dorsoventral axis of the *Drosophila* wing. See text for description of this network. (Adapted from Gerhart & Kirschner 1997. Reprinted by permission of Blackwell Science, Inc.)

expressing and not expressing *apterous* (Fig. 4b). Serrate protein is then secreted into the ventral compartment and associates with Delta protein to activate Notch receptor in the ventral cells (Fig. 4c). This stimulates symmetrical *wingless* expression (Fig. 4d), via *supressor of hairless* (Fig. 4e), in both the dorsal and ventral compartments of the imaginal wing disc. The boundary between these two compartments, where wingless protein is being secreted, becomes the margin of the wing blade, and is the major source of cell proliferation and growth of wing imaginal disc cells.

In comparing this regulatory network to that of the chick (Shubin et al 1997), one finds that both the chick *Lmx-1* and *Drosophila apterous* genes are expressed in the dorsal compartment of the developing wing imaginal disc and the dorsal compartment of the mesoderm in the developing limb bud. Both genes are necessary for specifying dorsal fate. As in *Drosophila*, the *Radical fringe* gene is expressed at the boundary of the dorsal and ventral compartments and activates the chick *Serrate-2* gene (Rodriguez-Esteban et al 1997). In turn, *Serrate-2* activates the *Notch–Delta* signalling pathway which leads to the formation of the AER, a structure analogous to the *Drosophila* wing margin where cell proliferation and growth takes place (Laufer et al 1997).

Based on these remarkable similarities, it is tempting to propose that these two regulatory networks are homologous. There are, however, several differences between these networks that support the hypothesis that they are only partly homologous. For example, *apterous* and *Lmx-1* may not be true orthologues of one another, and although they both play similar functional roles in specifying the dorsal fate in the developing *Drosophila* and chick wing, *apterous* has been shown to regulate *fringe* expression, whereas the chick *Lmx-1* gene does not (Irvine & Wieschaus 1994, Shubin et al 1997, Rodriguez-Esteban et al 1997). Furthermore, there appear to be novel functional linkages in the chick regulatory network that are absent from *Drosophila*. In *Drosophila*, wingless protein mediates patterning activity at the wing margin in the developing imaginal disc. Although the chick *Wnt7a* gene is expressed and required for the development of the AER, there is currently no evidence to suggest that any member of the vertebrate *wingless* gene family, including *Wnt7a*, possesses exactly the same function as the *Drosophila wingless* gene in the development of the AER (Brook et al 1996). Instead, patterning of the AER is carried out by two members of the fibroblast growth factor family, *Fgf4* and *Fgf8* (Cohn et al 1995, Crossley et al 1996). Interestingly, no members of the fibroblast growth factor family have been identified in *Drosophila* (Brook et al 1996).

Therefore, although there are many similar, perhaps homologous, elements in the regulatory networks that pattern the dorsoventral axis in *Drosophila* and chick limbs, there have also been evolutionary modifications and additions of elements, supporting the hypothesis that these networks are partly homologous. This example can be tentatively classified under scenarios 3 and 4 in Table 1, where the embryonic origins and morphological structures are convergent, but the gene regulatory network is partly homologous.

Prospects and challenges

Homology, one of the most crucial but complex biological concepts, can provide a framework to interpret comparisons between regulatory gene networks. As comparative data on regulatory gene networks accumulate, the similarity and phylogenetic criteria outlined in this chapter can serve as a methodology to distinguish whether regulatory networks are homologous, partly homologous or convergent. Setting up alternative hypotheses when interpreting comparisons for this class of data will advance our understanding of how regulatory gene networks evolve, as well as facilitate comparisons between the genetic networks of distantly related model organisms. Furthermore, since regulatory gene networks can be treated as a distinct level of biological organization, they can be incorporated into a hierarchical concept of homology. Perhaps one of the greatest challenges of the future will be to produce enough comparative data to analyse different levels of biological organization — such as genes, gene functions, gene networks, as well as embryonic origin and morphological structures — together in a phylogenetic framework. This will provide an integrated approach to the study of development and evolution.

Acknowledgements

I thank Alexa Bely, Chris Lowe, Margaret Pizer, Emile Zuckerkandl, Adam Wilkins and Greg Wray for insightful discussions and comments on earlier drafts of this manuscript. This work was supported by a graduate research fellowship from the Fonds pour la Formation de Chercheurs et l'Aide a la Recherche (FCAR) from the Quebec government.

References

Abouheif E 1997 Developmental genetics and homology: a hierarchical approach. Trends Ecol Evol 12:405–408

Abouheif E, Akam M, Dickinson WJ et al 1997 Homology and developmental genes. Trends Genet 13:432–433

Akimenko M-A, Ekker M, Wegner J, Lin W, Westerfield M 1994 Combinatorial expression of three zebrafish genes related to *Distalless*: part of a homeobox gene code for the head. J Neurosci 14:3475–3486

Arnone MI, Davidson EH 1997 The hardwiring of development: organization and function of genomic regulatory systems. Development 124:1851–1864

Brook WJ, Diaz-Benjumea FJ, Cohen SM 1996 Organizing spatial pattern in limb development. Annu Rev Cell Dev Biol 12:161–180

Carroll SB, Gates J, Keys DN et al 1994 Pattern formation and eyespot determination in butterfly wings. Science 265:109–114

Cohn MJ, Izpisùa-Belmonte JC, Abud H, Heath JK, Tickle C 1995 Fibroblast growth factors induce additional limb development from the flank of chick embryos. Cell 80:739–746

Crossley PH, Minowada G, MacArthur CA, Martin GR 1996 Roles for FGF8 in the induction, initiation, and maintenance of chick limb bud development. Cell 84:127–136

de Beer GR 1971 Homology: an unsolved problem. Oxford University Press, Oxford (Oxford Biol Readers 11)

De Robertis EM 1994 The homeobox in cell differentiation and evolution. In: Duboule D (ed) Guidebook to the homeobox genes. Oxford University Press, Oxford, p 11–24

Dickinson WJ 1995 Molecules and morphology: where's the homology? Trends Genet 11:119–121

Duboule D, Wilkins AS 1998 The evolution of 'bricolage'. Trends Genet 14:54–59

Gerhart J, Kirschner M 1997 Cells, embryos, and evolution. Blackwell Science, Boston, MA

Gilbert SF 1997 Developmental biology, 5th edn. Sinauer, Sunderland, MA

Goodrich LV, Johnson RL, Milenkovic L, McMahon JA, Scott MP 1996 Conservation of the *hedgehog/patched* signaling pathway from flies to mice: induction of a mouse *patched* gene by Hedgehog. Genes Dev 10:301–312

Gould SJ, Vrba ES 1982 Exaptation — a missing term in the science of form. Paleobiology 8:4–15

Greco TL, Takada S, Newhouse MM, McMahon JA, McMahon AP, Camper SA 1996 Analysis of the vestigial tail mutation demonstrates that *Wnt-3a* gene dosage regulates mouse axial development. Genes Dev 10:313–324

Hall BK (ed) 1994 Homology: the hierarchical basis of comparative biology, Academic Press, San Diego, CA
Holland PWH 1999 How gene duplication affects homology. In: Homology. Wiley, Chichester (Novartis Found Symp 222) p 226–242
Irvine KD, Wieschaus E 1994 *fringe*, a boundary-specific signaling molecule, mediates interactions between dorsal and ventral cells during Drosophila wing development. Cell 79:595–606
Kengaku M, Capdevila J, Rodriguez-Esteban C et al 1998 Distinct WNT pathways regulating AER formation and dorsoventral polarity in the chick limb bud. Science 280:1274–1277
Kitching L, Forey P, Humphries C, Williams D 1998 Cladistics: theory and practice. 2nd edn. Oxford University Press, Oxford
Lauder GV 1994 Homology, form, and function. In: Hall BK (ed) Homology: the hierarchical basis of comparative biology. Academic Press, San Diego, CA, p 152–189
Laufer E, Dahn R, Orozco OE et al 1997 Expression of *Radical fringe* in limb-bud ectoderm regulates apical ectodermal ridge formation. Nature 386: 366–373
Lowe CJ, Wray GA 1997 Radical alterations in the roles of homeobox genes during echinoderm evolution. Nature 389:718–721
Mayr E 1982 The growth of biological thought. Harvard University Press (Belknap), Cambridge, MA
Panganiban G, Irvine SM, Lowe C et al 1997 The origin and evolution of animal appendages. Proc Natl Acad Sci USA 94:5162–5166
Patterson C 1982 Morphological characters and homology. In: Joysey KA, Friday AE (eds) Problems of phylogenetic reconstruction. Academic Press, London (Systematics Association Special Volume series 21) p 21–74
Patterson C 1988 Homology in classical and molecular biology. Mol Biol Evol 5:603–625
Rodriguez-Esteban C, Schwabe JWR, De La Peña J, Foys B, Eshelman B, Belmonte JCI 1997 *Radical fringe* positions the apical ectodermal ridge at the dorsoventral boundary of the vertebrate limb. Nature 386:360–366
Roth VL 1988 The biological basis of homology. In: Humphries CJ (ed) Ontogeny and systematics. Columbia University Press, New York, p 1–26
Shubin NH 1994 History, ontogeny, and evolution of the archetype. In: Hall BK (ed) Homology: the hierarchical basis of comparative biology. Academic Press, San Diego, CA, p 250–267
Shubin N, Tabin C, Carroll S 1997 Fossils, genes, and the evolution of animal limbs. Nature 388:639–648
Wang J, Shackleford GM 1996 Murine *Wnt10a* and *Wnt10b*: cloning and expression in developing limbs, face and skin of embryos and in adults. Oncogene 13:1537–1544
Wray GA 1999 Evolutionary dissociations between homologous genes and homologous structures. In: Homology. Wiley, Chichester (Novartis Found Symp 222) p 189–206
Xue L, Noll M 1996 The functional conservation of proteins in evolutionary alleles and the dominant role of enhancers in evolution. EMBO J 15:3722–3731
Yuh C-H, Bolouri H, Davidson EH 1998 Genomic cis-regulatory logic: experimental and computational analysis of a sea urchin gene. Science 279:1896–1902
Zardoya R, Abouheif E, Meyer A 1996 Evolution and orthology of *hedgehog* genes. Trends Genet 12:496–497
Zuckerkandl E 1994 Molecular pathways to parallel evolution: I. gene nexuses and their morphological correlates. J Mol Evol 39:661–678
Zuckerkandl E 1997 Neutral and non-neutral mutations: the creative mix—evolution of complexity in gene interaction systems. J Mol Evol 44:s2–s8

DISCUSSION

Wagner: You seemed to be pessimistic about the ability to delineate networks, but your pessimism may be unwarranted. An approach similar to the one that Georg Striedter outlined yesterday could be used to look for co-variation of gene deployment in different organs and different species, and this would generate a variational definition of what usually goes together and what doesn't. This approach would avoid the arbitrariness of individuating a subnetwork from a large network.

Striedter: I also picked up on this point. It might be difficult to delineate gene networks in some cases, but it may often be possible. You said that the problems associated with homologizing gene networks are unique, but your diagrams are similar to neural networks. Neural networks suffer from the same problems you pointed out, namely that they may be hopelessly diffuse, but within this interconnected morass there may be some modules that are relatively individualized. For other analogies, you might also look toward metabolic control theory, where investigators are similarly struggling to identify functional modules within a complexly interconnected system (e.g. Rohwer et al 1996). Along these lines, what puzzled me about your presentation was that you seemed to be saying that networks can be individualized modules but that if they're not identical across species then they can't be 'fully' homologous. I don't understand how you can make these two points in the same chain of thought.

Abouheif: You have to be able to identify the same components. If some components have been added on, then you have some components that you can identify as homologous and some components as convergent. It's the mixing of these components that makes a network partly homologous.

Striedter: It sounds to me that you are requiring 'fully' homologous networks to be identical. But how would you permit a network to change during evolution?

Abouheif: Identical in what sense? It just has to have homologous genes, and these genes have to be doing similar things.

Wray: What you're saying is that this is similar to divergence, because divergence in morphology encompasses homologues that are not identical, but we don't use the term 'partial homology'.

Striedter: Yes. An evolving neural network can contract or expand, by co-opting other neural structures into itself. If I recognize networks at the network level of analysis, then I don't have to say that they're 'partially' homologous. They may be 'fully' homologous and simply change by incorporating new elements or deleting old ones.

Abouheif: But there's a difference between things that were present in the common ancestor and then diverged, and completely new things that have been added on after divergence. I'm trying to differentiate between those two situations.

Panchen: Is your definition of partial homology the same as that for transformational homology of structures?

Abouheif: No. At the structural level, the lower jaw of lizards and the ear ossicles of mammals are the same structures that have transformed. We recognize them because they have the same connections.

Panchen: But they're not the same.

Abouheif: They perform different functions, but it's the same elements that have transformed.

Panchen: That's transformational homology. Isn't that the same for your networks?

Abouheif: No. I would call it transformational homology if the same elements had transformed in one way or another. It is not transformational homology if new elements have just been added on.

Wagner: In the case of morphological characters, there are certain character transformations that follow inherent degrees of freedom, e.g. bones can become longer, shorter or can fuse, and there are others that lead to new constraints and new variational properties. The question is, can a similar way of thinking also be applied to these networks? For instance, are there intrinsic degrees of freedom in, say, rates of transcription, that can change without changing the basic structure of the network, and are there other properties that are more likely to be conserved, for instance as direct protein–protein interactions? If so, is it possible to sort out the intrinsic degrees of freedom for change from other properties that are intrinsically stable because of their biochemistry?

Abouheif: Once we understand more about the evolutionary properties we will be able to do this.

Wagner: The partial homology problem will then disappear if you think of homologues as being different states of the same element. It will be possible to distinguish between stages where some otherwise conservative feature is changed, and occasionally this will lead to the formation of new entities. This would be a way to distinguish between simple modifications and real novelties.

Hall: We want to be able to distinguish between the basic properties at each level. For example, you can tell a cartilage in the lower jaw of a reptile is the same as the cartilage in a middle ear because they share basic properties as cartilages, but they are completely different elements in terms of what they transform into.

Maynard Smith: I worry about this. One type of change in a gene network is the recruitment of a completely new gene with a different ancestry. One could recognize such an event by homology of gene sequences. It is not clear to me that if I was looking at a set of morphological structures, I would be able to recognize an analogous process in which a morphological structure had been altered by the recruitment of a new structural component. Genes are not like bones.

Wagner: One theory of the origin of the neocortex is that it is a composite structure composed of cell populations that in non-eutherian amniotes reside in different nuclei of the telencephalon and are integrated into the modules of the neocortex. This would be an example of several independent anatomical entities that had become integrated into something completely new, namely the neocortex.

Carroll: You didn't get as far as the phylogeny. Structures of the wings of the chick and *Drosophila* are not homologous. However, if there is a control network, then this must be partially homologous and have some other function. What would that other function for *engrailed*, *wnt*, *notch*, etc. be in the putative common ancestor or even in wing-less insects and dinosaurs?

Abouheif: This *wnt* gene pathway is not only involved in the setting up of the dorsoventral axis of the wing discs, but also the dorsoventral axis of the leg discs, the brain discs and the genital discs. Therefore, the system is used repeatedly in different discs.

Akam: There probably isn't a patterning process that doesn't use at least half of the genes in this pathway. What we are looking at is the homology of metazoan cells. Signalling is probably involved in every process where two cells become different from one another. We happen to have studied two developmental events that are similar, in terms of generating wings, and we have found that many of the same genes are involved.

Tautz: It is important to find out why particular cells carry out this signalling pathway in certain locations of the body. Therefore, we should look for the network of genes that cause the cells to be placed at this location.

Akam: To talk about, say, the Ras/GTPase signalling pathway as a significant developmental homology, a developmental network, seems nonsense. It is a bit of metazoan cell biology, and probably almost all cells have that pathway functioning at some level or another. In *Drosophila*, *dpp* expression in the wing is regulated by the *hedgehog/engrailed* signalling pathway in the third instar disc, but 10 hours later it is regulated by a completely different regulatory network. How can a pathway be conserved between the *Drosophila* wing and the chick wing, if it is not conserved 10 hours later in the *Drosophila* wing. In many of these comparisons we are dealing with complete nonsense.

Panchen: There is a word for it, i.e. analogy, and this doesn't commit you to any hypothesis.

Wagner: The confusion with analogy stems from the fact that in biology there are two definitions of 'function': (1) the biological role that the thing is adapted to do; and (2) what it is causally effective at doing.

Wray: When we deal with gene products, there is another distinction in the way people use the term 'function' in addition to that which you just described. For example, the biochemical function of a transcription factor is simply to bind to DNA, whereas its function in a developmental sense is defined by the specific piece of DNA to which it binds and the cell biological effects of that binding. Therefore, we need to be careful to distinguish which kinds of molecular 'function' we are talking about.

Rudolf Raff: Another complicating factor is that the networks which we find scattered throughout phylogeny have different downstream consequences, i.e. highly conserved networks can feed into different downstream functions and will produce the different consequences we observe. In this sense, 'function' has a different meaning.

Roth: Is there such a thing as homology or conservation of a role without the things playing that role actually being conserved?

Wray: We don't have enough data to make clear judgements in some of these cases. However, we have to be open to the possibility that this can occur, and we have to try to discriminate among alternative possibilities, or least keep an open mind about what the possibilities are.

Akam: The structure of promoter modules is relevant. Arnone & Davidson (1997) find that a typical promoter module involves about six transcription factors working together to give an output. No single site is critical, and the different transcription factors can be substituted for one another. Therefore, one can envisage a promoter module with a historical continuity of function, but the individual transcription factors that drive this module change rapidly in evolution, resulting in homologous modules that have less than half of their input factors in common.

References

Arnone MI, Davidson EH 1997 The hardwiring of development: organization and function of genomic regulatory systems. Development 124:1851–1864

Rohwer JM, Schuster S, Westerhoff HV 1996 How to recognize monofunctional units in a metabolic system. J Theor Biol 179:213–228

The effect of gene duplication on homology

Peter W. H. Holland

School of Animal and Microbial Sciences, The University of Reading, Whiteknights, Reading RG6 6AJ, UK

Abstract. Genes related by gene duplication within an organism's evolutionary lineage are termed paralogues; genes related by speciation are orthologues. It is generally agreed that orthologous genes must be compared when using DNA sequences to reconstruct the evolutionary history of organisms. There is an important exception: information from paralogous genes can reveal the root position of a phylogenetic tree. The duplicated rDNA genes of arrow worms provide an example. Gene duplication is also relevant when comparing gene expression between taxa; for example, when trying to identify homologous roles of genes. When gene duplication occurred after lineage divergence, single orthologues no longer exist, and comparison is complicated. This is a particular problem when comparing roles of vertebrate and invertebrate genes. Amphioxus and ascidian genes can be useful in such situations, since they diverged before extensive gene duplication in the vertebrate lineage. Using *Otx* and *Pax* as examples, I show how examination of amphioxus or ascidian genes reveals patterns of gene divergence after duplication, assisting the identification of homologous gene functions. Given the problems of comparing duplicated genes between species, the time is ripe for the introduction of additional terminology to elaborate on the concepts of paralogy and orthology.

1999 Homology. Wiley, Chichester (Novartis Foundation Symposium 222) p 226–242

The concept of homology can be applied to genes in the same way as it is applied to other types of character, although certain problems do arise. Distinguishing homology from convergent evolution is a practical problem, but there are also fundamental problems concerning the definition and use of homology as applied to genes. Simply to consider genes as homologous when they are descended from a common ancestral gene is accurate, but such a statement carries rather limited information when actually applied. For example, families of related genes often arise by gene duplication in evolution, such that several genes within one species can be homologous, and each may also be homologues to several genes in other species. A single term for both kinds of relationship does not convey sufficient information about the kind of evolutionary relationship, as first pointed out by

Walter Fitch. In a closing discussion to a paper on identifying homology, Fitch defined two subclasses of homology as applied to genes: paralogy and orthology (Fitch 1970). To quote:

> 'Where the homology is the result of duplication so that both copies have descended side by side during the history of an organism (for example, alpha and beta haemoglobin), the genes should be called paralogous (para = in parallel). Where the homology is the result of speciation so that the history of the gene reflects the history of the species (for example, alpha haemoglobin in man and mouse) the genes should be called orthologous (ortho = exact).'

These terms have become universally accepted by molecular evolutionary biologists, and have proved invaluable in discussions of homology at the molecular level. Here I examine three contexts in which gene duplication impinges on the concept and application of homology. First, I examine how gene duplication affects the construction of phylogenetic trees, with implications for assessment of morphological homology between taxa. Second, I examine how gene duplication complicates attempts to compare gene expression or gene function between taxa. A means to simplify the problem is proposed, based on careful selection of taxa uncomplicated by gene duplication. Third, I ask whether the current terms of homology, orthology and paralogy are adequate to cope with the complexity of evolutionary relationships between duplicated genes. I argue that new terms are required, and endorse some new terminology proposed by A. C. Sharman.

Gene duplication and phylogeny

Without phylogeny, we cannot determine whether similar morphological characters in different species are homologous. There has been optimism that so-called molecular phylogenetics will provide an objective and accurate means to reconstruct evolutionary trees. Gene duplication could potentially complicate this goal; it implies that simply using genes that are homologous between species is insufficient to ensure that the genes have followed the same evolutionary histories as the organisms that carry them. For example, alpha and beta globin genes are homologous, yet a molecular phylogeny constructed from several alpha globins and several beta globins may present an inaccurate view of species history. In the words of Fitch (1970), 'Phylogenies require orthologous, not paralogous, genes.'

One slight exception to this rule relates to the ribosomal DNA (rDNA) genes that are extensively duplicated, and exist in tandem arrays in animal genomes. Despite the fact that these genes are duplicated, they are extensively used in molecular phylogenetics. However, it is generally thought that this does not violate the rule of comparing orthologues, since genetic homogenization processes ensure that each duplicate copy does not accumulate mutations in isolation from the other copies. Mutations seem to be homogenized across the majority of rDNA genes within a species, by a range of genetic processes (Dover 1986). Since homogenization is thought to act faster than speciation, all rDNA genes within an organism are almost identical in DNA sequence, but different to those of related species. Thus, determining the DNA sequence of just one or a few rDNA genes per species should be sufficient to reveal species relationships, despite the fact that thousands of other rDNA copies may exist in each genome.

Max Telford and myself identified a case where rDNA sequences clearly violated the rule of comparing orthologues; even so, it yielded a positive outcome (Telford & Holland 1997). After cloning a region of the 28S rDNA gene from several species of chaetognath (arrow worms) by the PCR technique, we attempted to construct molecular phylogenies. As the first sequences were acquired, our initial attempts to produce phylogenies were so far at odds with accepted taxonomies that they were clearly in error. The source of the problem finally became apparent only after cloning multiple 28S rDNA genes from each species. We noted that many chaetognath species (probably all, although this was not proven) possess two distinct classes of 28S rDNA; we term these class I and class II sequences. When molecular phylogenetic analysis is applied to each class of sequence alone, internally consistent and realistic trees are produced; indeed, the class I and the class II trees had a similar overall structure. Evidently, genetic homogenization of rDNA genes is not occurring at a faster rate than speciation in chaetognaths. Instead, we conclude that an rDNA gene (or rDNA gene cluster) in an ancestor of all chaetognaths was duplicated; each daughter gene (or gene cluster) has been evolving in parallel with the organisms. Genetic homogenization is occurring within each rDNA class, but not between classes. In this example, sequencing of randomly chosen individual rDNA genes from each species produced a ludicrous phylogeny, since we were, in effect, comparing paralogous genes. It is not clear how widespread this phenomena is, but it does sound a warning note for rDNA phylogenetics.

Despite the warning above, the fact that paralogous rDNA genes were sequenced from chaetognaths did provide an advantage over a standard molecular phylogeny. Since they are descendent from a gene duplication, the two classes of rDNA yield trees that root each other (Fig. 1). This overcomes one of the notorious problems of molecular phylogenetics: determining directionality within the tree. Our two rooted trees clearly demonstrate that there are two clades within the phylum, equating to the previously recognized orders Aphragmophora and Phragmophora. This deep taxonomic split is therefore a true phylogenetic division.

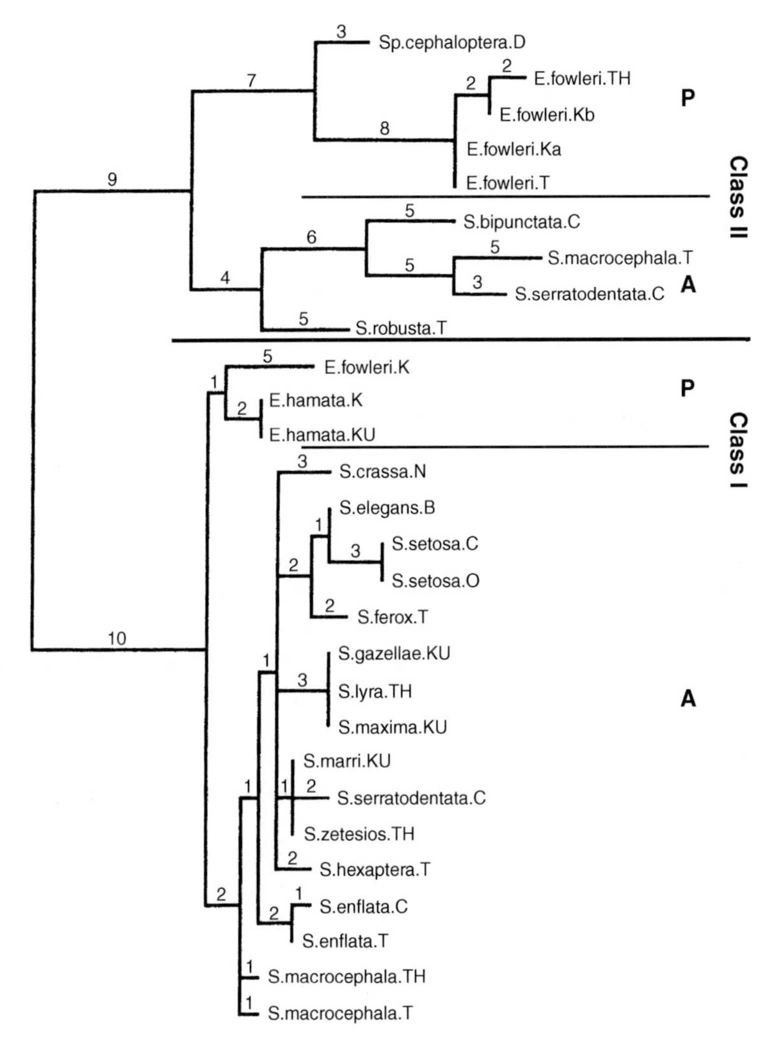

FIG. 1. The use of duplicated 28S rDNA genes allowed the root position to be determined for a molecular phylogeny of chaetognaths. The parallel trees each reveal the deepest phylogenetic division within the phylum, but should not be consulted for detailed aspects of the phylogeny, which were recovered from a separate sequence alignment. From Telford & Holland (1997). A, Aphragmophora; P, Phragmophora.

Rooting the chaetognath tree using duplicated rDNA genes is formally equivalent to a method previously used to propose a root for eukaryote, archaebacterial and eubacterial divergence, based on duplication of elongation factor or amino-acyl tRNA synthetase genes (Creti et al 1994, Brown & Doolittle

1995). In summary, duplicated genes should not always be avoided when constructing phylogenetic trees. As long as each paralogue is distinguished, duplicated genes provide an additional source of phylogenetic information, not easily extracted from orthologous genes alone.

Gene duplication and comparative developmental biology

There is a currently much enthusiasm to compare the expression patterns of homologous genes between representatives of divergent animal phyla. The fashion was founded upon the surprising discovery that spatial expression patterns were similar between homeotic genes of *Drosophila* and their homologues—the *Hox* genes—of mammals (Awgulewitsch et al 1986, Gaunt et al 1986, Duboule & Dollé 1989, Graham et al 1989). In the current phase of enthusiasm, however, similarities are not seen as surprises; conservation is now being hailed as the rule. For example, Gehring and colleagues have stressed the similarities in expression, and apparently similar role in regulating eye development, between a *Drosophila Pax* gene, *eyeless*, and a related gene in mammals, *Pax-6* (e.g. Halder et al 1995). Similarly, Bodmer, Harvey and others have stressed the similar expression, and apparently similar role in heart development, between a *Drosophila* NK gene, *tinman*, and a related gene in vertebrates, *NKx-2.5* (for review, Harvey 1986). Scott (1994) documents other examples.

This general approach of comparing the developmental roles of fly and vertebrate genes suffers from two quite different problems. First, the body plans of flies and vertebrates are so radically different that comparison of developmental roles is intrinsically difficult. Second, there is much evidence for extensive gene duplication during the early phase of vertebrate evolution (Holland et al 1994, Sharman & Holland 1996); hence, a given gene in an insect may be related to several genes in a vertebrate. In other words, directly orthologous genes cannot always be compared. These two compounding problems, body plan disparity and gene duplication, are present simultaneously. It would be helpful if these two problems were dissociated, so that the effects of each may be studied in isolation. One route to achieve this is to include data from the genes of amphioxus or ascidia. Since these animals are both chordates, they share many features of the body plan organization with vertebrates (e.g. notochord, dorsal nerve cord, branchial region, musculature derived from paraxial mesoderm). They also diverged from the chordate lineage before the extensive gene duplications that characterize the early phase of vertebrate evolution (Holland 1996). There is a higher chance, therefore, that they will possess direct orthologues of particular insect genes, although this cannot be taken for granted. Comparison of gene expression patterns between amphioxus and vertebrates, therefore, may be useful in deducing the consequence of gene duplication for gene expression or gene function, within the framework of a similar body plan (Fig. 2). Comparison between amphioxus and *Drosophila* may remove the complication of gene duplication, and simplify attempts to compare gene expression or gene function between divergent body plans (Fig. 2).

Two examples will illustrate this approach: the *otd/Otx* and the *Pax-3/7* gene families. The *orthodenticle* (*otd*) homeobox gene of *Drosophila* is one of three gap genes with roles in developmental patterning of the fly head; later in development the gene also has specific roles within the developing fly brain (Finkelstein et al 1990). Two homologues have been cloned from mouse, *Otx1* and *Otx2*; two or three homologues have also been cloned from several other vertebrates. Since the expression patterns of (at least some) *Otx* genes are anterior, reminiscent of the *Drosophila otd* gene, it has been speculated that genes of the *otd/Otx* gene family have an evolutionarily conserved role in head development in flies and mice (for review, see Holland et al 1992).

We recently described the cloning of a member of the *otd/Otx* gene family from amphioxus (Williams & Holland 1996, 1998). Southern hybridization strongly indicated that this gene, *AmphiOtx*, is the only member of the *otd/Otx* gene family in the amphioxus genome; a conclusion also supported by our isolation of the same gene from a genomic library, cDNA library and by PCR. We conclude therefore that *AmphiOtx* is the direct orthologue of *Drosophila otd*; a statement that

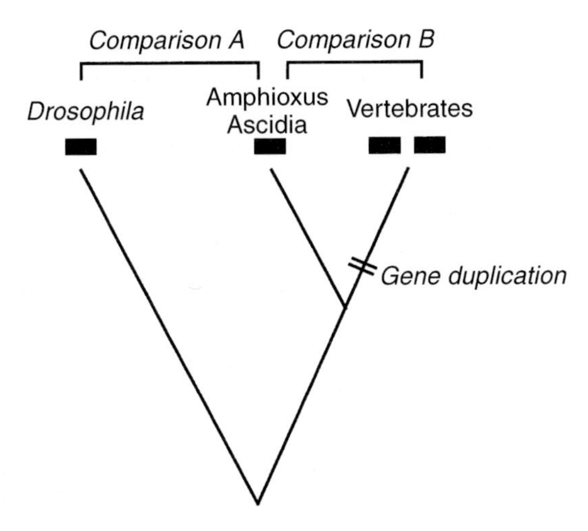

FIG. 2. Inclusion of ascidian or amphioxus genes can assist comparison between arthropod and vertebrate genes. Comparison A allows genes to be compared between phyla without the complication of vertebrate gene duplications; comparison B allows the effect of gene duplication to be studied without the complication of disparate body plans.

cannot be made for either *Otx1* or *Otx2* of mouse. With the complication of gene duplication removed, how does the expression of a chordate *Otx* gene compare with that of an arthropod *otd* gene? During early development *AmphiOtx* is expressed in a broad region of the anterior part of the embryo; first in endoderm, and then in overlying neurectoderm (Williams & Holland 1996). Expression later focuses down to specific cells in the cerebral vesicle, including putative photoreceptors and the cells secreting Reissner's fibre. The extensive early expression is similar to the broad stripe of expression described for *Drosophila otd* in the blastoderm stage embryo. The data are consistent with, but not proof of, the suggestion that both genes play a conserved role in head development: a homologous developmental role.

Comparison between amphioxus and vertebrates allows us to deduce the impact of gene duplication on gene structure and expression, if (and only if) the amphioxus *AmphiOtx* gene is descendant from the precursor of the multiple genes in vertebrates. Two sources of evidence independently demonstrate that this is the case (Williams & Holland 1998). First, all vertebrate *Otx* genes have a tandem duplication of a C-terminal domain of the protein, the 'Otx-tail'. In contrast, amphioxus has a single domain, representing the ancestral condition. Second, molecular phylogenetic analysis clearly places *AmphiOtx* as an outgroup to all vertebrate *Otx* genes (Fig. 3). The latter genes split clearly into two groups: the *Otx2* gene of several species, and a group containing *Otx1* and duplicates of it in *Xenopus* and zebrafish. A second conclusion apparent from the phylogenetic tree is that the rate of molecular evolution has been far greater in the *Otx1*-like genes than in the relatively conservative *Otx2* genes. A parallel conclusion emerges from expression and function studies. The mouse *Otx2* gene is expressed broadly at the anterior end of the early embryo, only later focusing down to the forebrain and midbrain, and finally to specific cell types in the brain; this spatiotemporal pattern is reminiscent of *AmphiOtx* described above. Gene targeting reveals an important role in early head patterning, since *Otx2* mutant mice lack a large portion of the anterior head region (Ang et al 1996). In contrast, mouse *Otx1* is more accurately considered as a forebrain- and midbrain-specific gene; its mutant has a much milder phenotype, with relatively minor neurological alterations (Acampora et al 1996). In summary, the more constrained gene (*Otx2*) has retained an expression pattern comparable to the inferred ancestral pattern, whereas *Otx1* has diverged to acquire novel roles concerned with the detailed architecture of the brain. This pattern of molecular evolution fits perfectly with the model proposed by Ohno (1970). In his words: 'By duplication, a redundant copy of a locus is created. Natural selection often ignores such a redundant copy, and, while being ignored, it accumulates formerly forbidden mutations and is reborn as a new gene locus with a hitherto non-existent function.'

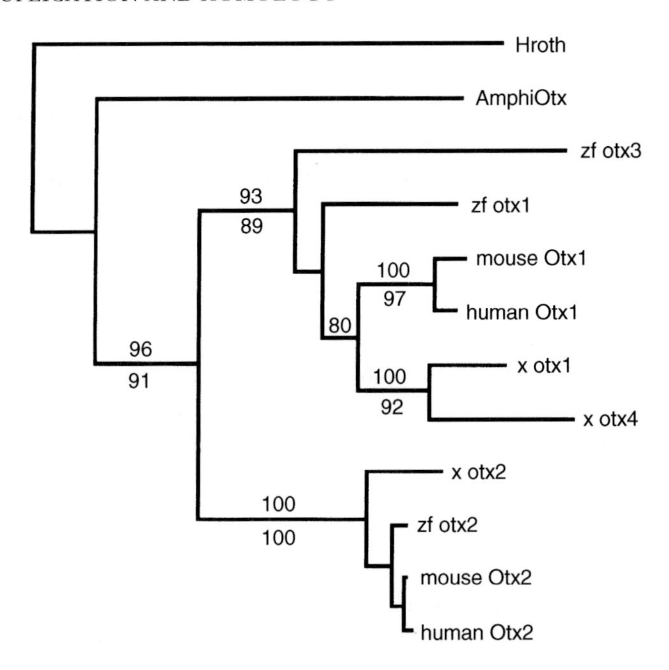

FIG. 3. Molecular phylogeny of the *otd/Otx* gene family, showing amphioxus *AmphiOtx* as descendent from the precursors of *Otx2*- and *Otx1*-like genes, and the different rates of evolution of the latter two genes. From Williams & Holland (1998).

The second example I will discuss demonstrates that this simple model does not always hold. The *Pax* multigene family includes at least nine genes in the mouse and eight in *Drosophila*. Molecular phylogenetic analysis divides *Pax* genes into four or five subfamilies (Wada et al 1998). One includes vertebrate *Pax-3* and *Pax-7* plus three *Drosophila* genes: *prd* (a pair-rule gene), *gsb* (a segment polarity gene) and *gsb-n*. An ascidian member has been cloned, *HrPax-37* (Wada et al 1996, 1997). The ascidian sequence is an outgroup to the two vertebrate homologues, demonstrating that a gene duplication occurred in the vertebrate lineage. Comparison between ascidian and vertebrates, therefore, can be used to examine the consequence of gene duplication, within a similar body plan. *HrPax-37* functions in dorsalization of the neural tube (Wada et al 1997), and is also expressed segmentally in spinal cord (Wada et al 1996). *Pax-3* and *Pax-7* have roles comparable to the first of these (in neural tube closure and, for *Pax-3*, in neural crest cell specification), but both also have an additional role in specification of dorsal mesodermal fates. Neither *Pax-3* nor *Pax-7* show evidence of a role in segmentation. Hence, in this example, gene duplication was probably associated with divergence of both descendent genes; this is a different course of evolution from the *Otx* genes. I argued above that ascidian or amphioxus genes may often be more suitable than vertebrate genes to compare to *Drosophila*, since gene duplications in vertebrates complicate the comparison. This does not hold for this gene family, however, since molecular phylogenetic analyses reveal that gene duplications have also occurred in the lineage leading to *Drosophila* (to yield *prd*, *gsb* and *gsb-n*).

Gene duplication: new terminology

In the above discussions, many cases of homology between genes did not fall neatly into the alternatives of paralogy or orthology proposed by Fitch (1970). There are at least two types of evolutionary relationship between duplicated genes to which this applies. The first is when gene duplication occurs in the lineage of one species, but not in a sister lineage. For example, amphioxus *AmphiOtx* is neither paralogous nor orthologous to mouse *Otx1* (or *Otx2*). Similarly, ascidian *HrPax-37* is neither paralogous nor orthologous to mouse *Pax-3* (or *Pax-7*). A. C. Sharman (personal communication 1998) proposes two new terms to describe these relationships; two are required since this is a directional relationship (like father and son). 'Pro-orthologous' describes the relationship of a gene to one of the post-duplication descendants of its orthologue. For example, *AmphiOtx* is the pro-orthologue of *Otx1* (or of *Otx2*); *HrPax-37* is the pro-orthologue of *Pax-3* (or of *Pax-7*); the nematode 28S rDNA gene cluster is pro-orthologous to the class I (or class II) rDNA genes of chaetognaths. 'Semi-orthologous' describes the converse: the relationship of one of a set of duplicated genes to a gene directly descendent from the ancestor of the whole set. The second case when a new term is required is when gene duplication has occurred independently in two lineages. Only one term is needed, since the relationship has no directionality (like cousins). Sharman proposes the term 'trans-homologous' for this relationship. Thus, *Drosophila prd* is the trans-homologue of mouse *Pax-3*. The three new terms are supplements, not replacements, to the terms orthology and paralogy; the latter have proved their worth and will continue do so in discussions of molecular homology.

Summary

(1) When a gene duplicates in evolution, it will have homologues within the same species. Fitch (1970) proposed the term 'paralogous' to describe this homology. The term 'orthologous' was proposed for precisely homologous genes in different species.

(2) It is widely accepted that orthologous genes must be used when constructing phylogenetic trees. Using the of 28S rDNA genes in arrow worms, I show

that paralogous genes should not be neglected in molecular phylogenetics, since they can provide insight into root position.

(3) Attempts to compare developmental expression or function of genes between arthropods and vertebrates are complicated by two factors simultaneously: body plan disparity and the effect of gene duplication in vertebrates. Inclusion of ascidian or amphioxus genes may allow these factors to be considered independently.

(4) In the *otd/Otx* gene family, inclusion of amphioxus suggested that after gene duplication in vertebrates, one daughter gene retained an ancestral role, whilst a second daughter gene diverged. In the *Pax-3/7* gene subfamily, inclusion of ascidian suggested that after gene duplication in vertebrates, both daughter genes diverged.

(5) The terms paralogous and orthologous are insufficient to describe all homologous relationships between genes. A singleton gene is pro-orthologous to a duplicate of its orthologue in another lineage; the reverse relationship is semi-orthology. Independent gene duplication in two lineages results in trans-homology.

Acknowledgements

I thank Max Telford, Nic Williams and Hiroshi Wada for skilfully performing the experiments on which these conclusions are based. My conviction that new terminology is required for molecular homology dates back to 1994, but was refined more recently through discussions with Anna Sharman; the terms themselves were proposed by Anna Sharman, incorporating lexicographic suggestions from Jose Campos-Ortega. Our work on rDNA genes was funded by Natural Environment Research Council, on *Otx* by Biotechnology and Biological Sciences Research Council and on *Pax* by Human Frontiers Science Program.

References

Acampora D, Mazan S, Avantaggiato V et al 1996 Epilepsy and brain abnormalities in mice lacking the *Otx1* gene. Nat Genet 14:218–222

Ang S-L, Jin O, Rhinn M, Daigle N, Stevenson L, Rossant J 1996 A targeted mouse *Otx2* mutation leads to severe defects in gastrulation and formation of axial mesoderm and to deletion of the rostral brain. Development 122:243–252

Awgulewitsch A, Utset MF, Hart CP, McGinnis W, Ruddle FH 1986 Spatial restriction in expression of a mouse homeobox locus in the central nervous system. Nature 320:328–335

Brown JR, Doolittle WF 1995 Root of the universal tree of life based on aminoacyl-tRNA synthetase gene duplications. Proc Natl Acad Sci USA 92:2441–2445

Creti R, Ceccarelli E, Bocchetta M et al 1994 Evolution of translational elongation factor (EF) sequences: reliability of global phylogenies inferred from EF-1alpha (Tu) and EF-2(G) proteins. Proc Natl Acad Sci USA 91:3255–3259

Dover GA 1986 Molecular drive in multigene families: how biological novelties arise spread and are assimilated. Trends Genet 2:159–165

Duboule D, Dollé P 1989 The structural and functional organization of the murine *Hox* gene family resembles that of *Drosophila* homeotic genes. EMBO J 8:1497–1505
Finkelstein R, Smouse D, Capaci TM, Spradling AC, Perrimon N 1990 The *orthodenticle* gene encodes a homeodomain protein involved in the development of the *Drosophila* nervous system and the ocellar visual structures. Genes Dev 4:1516–1527
Fitch W 1970 Distinguishing homologous from analogous proteins. Syst Zoology 19:99–113
Gaunt SJ, Miller JR, Powell DJ, Duboule D 1986 Homeobox gene expression in mouse embryos varies with position by the primitive streak stage. Nature 324:662–664
Graham A, Papalopulu N, Krumlauf R 1989 The murine and Drosophila homeobox gene complexes have common features of organization and expression. Cell 57:367–378
Halder G, Callaerts P, Gehring WJ 1995 Induction of ectopic eyes by targeted expression of the *eyeless* gene in *Drosophila*. Science 267:1788–1792
Harvey RP 1996 NK-2 homeobox genes and heart development. Dev Biol 178:203–216
Holland PWH 1996 Molecular biology of lancelets: insights into development and evolution. Isr J Zool 42:S247–S272
Holland PWH, Ingham PW, Krauss S 1992 Mice and flies head to head. Nature 358: 627–628
Holland PWH, Garcia-Fernàndez J, Williams NA, Sidow A 1994 Gene duplications and the origins of vertebrate development. Development 1994 Suppl 125–133
Ohno S 1970 Evolution by gene duplication. Springer-Verlag, Heidelberg
Scott MP 1994 Intimations of a creature. Cell 79:1121–1124
Sharman AC, Holland PWH 1996 Conservation, duplication, and divergence of developmental genes during chordate evolution. Neth J Zool 46:47–67
Telford MJ, Holland PWH 1997 Evolution of 28S ribosomal DNA in chaetognaths: duplicate genes and molecular phylogeny. J Mol Evol 44:135–144
Wada H, Holland PWH, Satoh N 1996 Origin of patterning in neural tubes. Nature 384:123
Wada H, Holland PWH, Sato S, Yamamoto H, Satoh N 1997 Neural tube is partially dorsalized by overexpression of *HrPax-37*: the ascidian homologue of *Pax-3* and *Pax-7*. Dev Biol 187:240–252
Wada H, Saiga H, Satoh N, Holland PWH 1998 Tripartite organization of the ancestral chordate brain and the antiquity of placodes: insights from ascidian *Pax-2/5/8*, *Hox* and *Otx* genes. Development 125:1113–1122
Williams NA, Holland PWH 1996 Old head on young shoulders. Nature 383:490
Williams NA, Holland PWH 1998 Gene and domain duplication in the chordate *Otx* gene family: insights from amphioxus *Otx*. Mol Biol Evol 15:600–607

DISCUSSION

Wray: There is another consequence of gene duplication besides the two scenarios you outlined. This is exemplified by *distalless*, where a set of roles has been partitioned between the two products of the duplication.

Peter Holland: This is probably going to turn out to be common. It's easy to envisage why this occurs. If a gene is pleiotropic it can't be optimized for all the functions, but if it is duplicated then the two genes can become specialized.

Wilkins: You compared the coding sequence of *Otx1* and *Otx2* but have you or anyone else also looked at the flanking *cis* regulatory sequences?

Peter Holland: We haven't. It would be interesting to do, although it may be difficult to interpret because of the time-scales that we're dealing with since gene duplication.

Meyer: In relation to time-scales and levels of comparison, we should be aware that in the comparisons of developmental genes and mechanisms in *Drosophila* and mouse, there is a large phylogenetic gap of hundreds of millions of years. Therefore, if we want to understand gene duplication events, network patterns or changes in function it may be more beneficial to look at a closely related set of species, e.g. within the *Drosophila* genus, because although the gene function will not be significantly different, focusing on the processes at this 'fine-grained' evolutionary level might be useful.

Peter Holland: It would be useful to do this, but we shouldn't avoid making large comparisons as well because these are an interesting aspect of phylogeny.

Rudolf Raff: It may be important to look at basal taxa because the organisms that have been used for these studies, e.g. *Drosophila*, are often highly modified.

Wray: It's also worth pointing out that the most basally branching members of the clade don't necessarily retain a true picture of what the ancestor of that clade looked like.

Tautz: We are currently comparing the *Drosophila* segmentation gene network in other organisms at the molecular level, and at least at this level, it does not seem to be so dramatically different. Thus, even though *Drosophila* shows a highly derived form of development, the basic processes may not be so different from organisms which show a less derived form.

Wray: But this does highlight the importance of intensive phylogenetic sampling. One of the important implications of Peter Holland's talk is that if you do point-to-point comparisons between a few taxa, and you miss out the intermediate steps, you are missing out important information. By filling in intervening taxa, you obtain a much richer amount of information.

Rudolf Raff: It's not just extra taxa. Peter Holland has chosen a basal type of chordate to study, and he's eliminated the serious problem of genomic duplication events in vertebrate evolution.

Panchen: You can produce hypothetical ancestors by producing a cladogram if you have enough different species.

Peter Holland: I agree, but it is not easy to decide which is a basal taxon because you can rotate any node in the cladogram. For the character you are interested in, you have to find the organism that has the plesiomorphic condition. There are genes within amphioxus that have undergone duplications, and for those genes amphioxus is not a good species in which to do these comparisons.

Elizabeth Raff: Another example is the elegant work by Piatigorsky (1998), Wistow (1993) and their colleagues on the recruitment of metabolic enzymes to function as lens crystallins. They have examples both of gene duplication preceding change of function, and change of function preceding gene duplication.

Lacalli: Thinking about the evolutionary consequences of gene duplication in a simple way, I would have predicted that morphological diversification would follow duplication events, but amphioxus doesn't seem to bear this out. If you compare amphioxus and other invertebrates, the striking thing is how many vertebrate-like features amphioxus has that are not found elsewhere among invertebrates. So the evolutionary novelty is there already in rudimentary form before duplication, although it is elaborated further after duplication.

Peter Holland: The key phrase in that statement is 'in a rudimentary form'.

Wray: It's also possible to have diversification of function without gene duplication. Therefore, this is an important mechanism, but it's not the only mechanism.

Lacalli: But there seems also be some kind of limit on what can be achieved without the extra genes. The endostyle in ascidians, for example, combines various hormone-secreting and feeding functions that are distributed among various subsidiary organs in vertebrates. The ascidian organ seems to be required to perform a mixed bag of functions, as if it were blocked from further diversification.

Akam: It is perhaps a weak argument to say that genes must be duplicated to increase regulatory complexity. We should be thinking in terms of life history strategies and pressure to keep genomes compact. It may not matter to most vertebrates how much DNA they have and how many genes they have. Clearly, in some lineages they have reached the point where there are vast excesses of DNA over what might be required to build a vertebrate body plan, yet it doesn't seem to have made them more complex vertebrates.

Peter Holland: When we looked at the evolution of genes encoding enzymes, we found examples of gene families in which all the duplicated genes have been lost. In contrast, in the case of homeobox genes, for example, the duplicated genes are more often retained. A pattern is therefore emerging which suggests that the duplication of developmental genes does allow additional potential to be realized.

Wilkins: Is polyploidation a good thing in terms of promoting innovation or, in contrast, does it promote stabilization? In all the old textbooks polyploidation is described as a stabilizing event, but there seems to have been a change in the way people look at polyploidation.

Peter Holland: I suspect it depends on the context. In some lineages polyploidization may promote stabilization, and in other lineages it may promote additional complexities. It probably depends on the genome, the environment and what constraints there are on gene number and diversity.

Maynard Smith: Most of the increase in DNA content between invertebrates and bacteria has arisen by duplication of bits of genome of various sizes. An invertebrate could not be made with the amount of DNA there is in *Escherichia coli*, so ultimately an increase in DNA by duplication has been required for evolutionary complexity. It may not always be used that way, because clearly there are examples of duplication in which the organism does nothing with the extra copies.

Müller: I agree, but just because gene duplication seems to be a requirement for morphological evolution we should probably not conclude that this kind of genetic evolution is causal in driving morphological evolution. We should be aware of the epigenetic contexts in which these gene duplications are taking place in order to understand how genomic evolution possibly co-opts functions of the epigenetic systems that were there already.

Akam: I look at the same data in the opposite way. I'm staggered that it is possible to make a nematode with only fourfold the amount of genes as there are in *E. coli*. We're not talking about large increases in genome size.

Maynard Smith: At the moment we do not have a way of assessing how much genetic information is required to make a structure with a certain level of complexity. However, there is still a crude relationship between the amount of DNA and complexity, although doubling the amount of DNA is not going to add any relevant information until the new copy is altered and programmed by selection.

Wagner: One may consider that an increased amount of DNA has direct physiological consequences that may be directly adaptive, e.g. DNA amount influences cell size and metabolic rate, and it affects cell cycle length. Those direct physiological consequences may drive the evolution of genome size at a microevolutionary level.

Striedter: How often is it the case that following gene duplication there has been so much divergence that the duplication is now beyond our level of detectability? Could this result in parallelism? This line of thought also leads me to wonder whether all new genes in some sense have arisen as a result of duplications.

Peter Holland: It's not possible to put any numbers on this. We would all agree that genes can diverge beyond the level at which we would conclude they are from the same gene family. Falciani et al (1996) have published an example of this, i.e. the *zen* gene of *Drosophila,* which has been shown to have been derived from a *Hox* gene. Taken out of context, the sequence of *zen* looks so different from a *Hox* class that its origin could not have been deduced. Data from additional taxa had to be incorporated to demonstrate this.

Rudolf Raff: On a larger scale, however, there are only a limited number of major gene families, or superfamilies, which indicates that over time within evolutionary history there have been repeated duplications, but that these still retain relationships, so that one can go back and identify gene families and superfamilies.

Maynard Smith: It does look as though almost all coding DNA has arisen by the duplication of DNA that was previously coding for something. There is a mechanism, however, whereby something quite different could happen, namely the occurrence of a frameshift mutation, which from a functional point of view gives rise to a gene coding for a completely unselected random sequence of amino acids. I'm not aware of any evidence that this has ever been used in evolution to produce a new functional protein with a new adaptation, but it is possible that it has. It would be interesting to know whether it is possible to go anywhere with a completely new random sequence.

Tautz: There is currently too much focus on duplicated genes and modules that have been duplicated and swapped between genes. We have started to look at genes that do not have modules, and about one-third of the eukaryotic genome does not have modules that are recognized in other organisms. These tend to evolve rapidly (Schmid & Tautz 1997) and therefore are responsible for a lot of diversity, but we don't know whether they have evolved from scratch from random sequences and we don't know what they do.

Meyer: Another mechanism is exon shuffling, which might occur early in evolution but would still give rise to variability.

Hinchliffe: There has been surprisingly little mention of neoteny at this symposium. I wondered how far you felt the old Garstang (1928) theory of the origin of the chordates by neoteny from an ascidian-like tadpole is supported by the molecular evidence?

Peter Holland: I don't think it is well supported at all. Phylogenies of the urochordates suggest that forms such as appendicularians are basal urochordates, which does not fit with Garstang's scheme.

Lacalli: There are real problems in deciding where the ascidians fall in the tunicate lineage, and this is related to the question of whether the ancestral form was basically pelagic or sedentary. If the appendicularia are basal, the case for the pelagic habit being ancestral is much stronger. Yet modern appendicularians are specialized organisms, which suggests that lots of secondary modification may have occurred. I would also argue that ascidians are divergent, and that pelagic forms like the salps may be closer to the ancestral form. I base this partly on CNS structure, because ascidians show a much greater degree of reduction and asymmetry in both the larval and adult CNS, implying that lots of secondary alterations have occurred. So I would suggest that even the ascidian tadpole larva, which is often taken as a model for the ancestral form, may be quite modified.

Peter Holland: The sequences of the ascidian *Hox* genes we've looked at are highly divergent.

Nicholas Holland: Cephalochordates are the sister group of the vertebrates, whereas ascidians do not even appear to be the most basal group of tunicates. Brooks, in the last century, believed that the ancestral tunicate (and thus chordate) was like an appendicularian, and not like an ascidian. This appendicularian theory of chordate origins was revived by Linda Holland et al (1988), who found that an appendicularian sperm had an acrosomal granule, which underwent the acrosome reaction and harpooned the egg. The egg contained cortical granules that were ejected and participated in fertilization envelope formation. It is controversial whether ascidian sperm have acrosomes, and their eggs certainly don't contain cortical granules. Therefore, if you look at Garstang's tree, you have to get rid of several important things that are clearly there in echinoderms and enteropneusts, and reinvent them in appendicularians. It was gratifying some years later when Wada & Satoh (1994) did the rDNA work and found that appendicularians were the most basal group of tunicates.

Wilkins: To go back for a moment to more general issues, I have been wondering why the debate about homology didn't get going until sometime after Darwin. My impression is that Owen provided a definition of homology, but that this was an essentialist and typological explanation. Darwin then put this in an evolutionary context, and at the time this was seen as a great advance. The concept that things are the same regardless of form or function because they are related by descent was, at the time, useful because Darwin was talking about organisms that were obviously related. The debates about whether specific structures are homologous started when people began to look at deeper divergences. We can look at the problem this way: imagine you are tracing a particular composite trait back through its various lineages in evolutionary time. What you would find is that as you go further and further back, the trait looks less and less similar to the trait present today. Most of the debates that have taken place about whether a particular structure is homologous in two different species occur when much of the similarity has been lost. If you find yourself in such a debate, I suggest that you are wasting your time, and you should either try to obtain more data, if this is possible, so that you can resolve it, or you should drop it and go on to discuss things that can be resolved.

Tautz: I disagree. The idea of the archetype is that it represents a common element that has remained common even after a long period. It is necessary to define this archetype, but not to think about other elements that have diverged.

Wilkins: That is a typological approach.

Wray: It refers to conservation and not to an archetype.

Panchen: A typological approach conflates two different concepts: first, idealism, where there's an archetype; and second, essentialism, where everything is based on definition.

Wagner: Adam Wilkins' point is important because there are phases of stasis with respect to some of the characters, and it is interesting to think about what happens when these phases of stasis are broken.

Wilkins: For traits that remain the same over long periods in different lineages, one doesn't debate whether they're homologous. There is general agreement. The problems arise, as in the case of the neocortex in birds and mammals, when there has been a lot of change.

Wray: An interesting parallel was pointed out by Joseph Felsenstein (1978), i.e. in particular phylogenetic trees where convergence is tied to branch length there are regions in which it is difficult to recover history. Systematists now call this the 'Felsenstein zone' (Swofford et al 1996). I propose that we call the kind of situation Adam Wilkins just described the 'Wilkins zone'!

Hall: This is clearly the essence of the problem. We are trying to talk about similarity and sameness in the context of 500 million years of evolution.

Greene: About half way through this symposium I began to feel like I did throughout the adaptation controversies of the 1980s, after which I decided I could continue with my work without having to use the word adaptation. So I wondered if I could do the same with the word homology. What I do is map characters onto phylogenies and make inferences about transitions, so it seems like the best thing for me to do is to talk about things as taxic homologues, and then there would be no ambiguity about what I was doing, and that when I wanted to think about underlying phenomena I would talk about biological homologues.

Hall: That's one of the points I will raise in my summary, i.e. in some cases we could probably do quite well without using the term homology.

References

Falciani F, Hausdorf B, Schröder R et al 1996 Class 3 *Hox* genes in insects and the origin of *zen*. Proc Natl Acad Sci USA 93:8479–8484

Felsenstein J 1978 Cases in which parsimony and compatibility methods will be positively misleading. Syst Biol 27:401–410

Garstang W 1928 The morphology of the Tunicata and its bearing on the phylogeny of the Chordata. Q J Micros Sci 72:51–187

Holland LZ, Gorsky G, Fenaux R 1988 Fertilization in *Oikopleura dioica* (Tunicata, Appendicularia): acrosome reaction, cortical reactivation and sperm–egg fusion. Zoomorphology (Berl) 108:229–243

Piatigorsky J 1998 Multifunctional lens crystallins and corneal enzymes. More than meets the eye. Ann NY Acad Sci 842:7–15

Schmid JK, Tautz D 1997 A screen for fast evolving genes from *Drosophila*. Proc Natl Acad Sci USA 94:9746–9750

Swofford D, Olsen GJ, Waddell PJ, Hillis DM 1996 Phylogenetic inference. In: Hillis DM, Moritz C, Mable BK (eds) Molecular systematics, 2nd edn. Sinauer Associates, Sunderland, MA, p 407–514

Wada H, Satoh N 1994 Details of the evolutionary history from invertebrates to vertebrates, as deduced from the sequence of 18S rDNA. Proc Natl Acad Sci USA 91:1801–1804

Wistow G 1993 Lens crystallins: gene recruitment and evolutionary dynamism. Trends Biochem Sci 18:301–306

Summary

Brian Hall

Department of Biology, Dalhousie University, Halifax, Nova Scotia, Canada B3H 4J1

I started this symposium with four signposts that I thought may be useful: (1) what is homology; (2) what is it not; (3) levels of homology; and (4) research programmes. And we touched on all of those points to some extent.

Many textbooks of evolution do not define evolution or they have multiple definitions, because evolution operates on many levels. That doesn't worry us. We read textbooks on evolution even though many don't have explicit definitions of evolution. Homology is tied to evolution—because organisms, structures and genes evolve—even though it was originally defined outside an evolutionary context. Homology, like evolution, is going to operate at different levels. One of the points that has come out of this symposium is that homology is hierarchical; it operates at the phenotypic, structural and genetic levels. Again, this is not something that worries us because it seems inappropriate to reduce homology to a particular level or to a particular definition. David Wake started us off on this train of thought by saying let's agree that homology exists and let's move on with it. Part of this process of moving on with it involves coming up with sub-definitions, such as those we have heard about in Peter Holland's presentation, and the remainder involves the study of evolution and the organization of life. The challenge is to find the links between the levels, just as this is one of the major challenges in evolution.

What is homology not? The word analogy has been mentioned rarely, so it seems as though we're not concerned with homology and analogy being opposites. If homology is tied to evolution, then the opposite of homology is something that's not tied to evolution, which is something we have not been concerned about. The issue is the criteria we should use to determine whether something is not homologous. Such criteria may include the answer to the question, how deep is the shared ancestor? If that ancestor is deep, we are not talking about homology, but rather we are probably talking about homoplasy, convergence or parallelism. When trying to answer this question, we must also bear in mind what we mean by sameness (i.e. when we talk about homologous structures being the same), what we mean by continuity of information and what we mean by conservation. I'm not sure we have resolved these issues. Some of us feel that homology and homoplasy are similar, whereas others feel they are different or that homoplasy represents deep homology.

I also want to mention the notion of levels, and illustrate this with a specific example. A structural morphologist looks at the ends of the anterior appendages on three different vertebrates and decides that as hands these structures are homologous. They then look in more detail and find that there are arrays of elements at the end of these structures, i.e. digits, and they conclude that they are also homologous. They may even be prepared to say that because there are five digits on each of those hands, it is possible to homologize individual digits. Therefore, at the level of the phenotype—as a hand and as a set of digits—these are homologous. Consider, then, three developmental biologists, who look at the processes underlying the development of these three homologous structures. They decide that the appendages of two of the vertebrates use the same developmental processes, but the other does not. They, however, come to three different conclusions. The first concludes that homologous structures can arise by different developmental processes, i.e. development evolves, and it is not possible to make any further statements. The second concludes that development evolves, but the structures are not homologous as hands or as digits because the developmental processes are not homologous as developmental processes. Clearly, there are now difficulties because homology is being translated from one level to another, and it may be inappropriate to draw conclusions from one level to another. The third concludes that development evolves, but that the structures are still homologous, i.e. there are two different levels of homology, and even though the developmental processes are divergent, it is still possible to identify homology at a structural level.

Consider now three geneticists, who begin to analyse the genes involved in controlling these developmental processes. They look into whether the same genes are involved, whether those genes have similar sequences, whether the same genetic networks are involved and whether the genes are expressed in the same places. They discover that two of the vertebrates are homologous at the genetic level, but that these two vertebrates do not share the same developmental pathway. The third vertebrate is not homologous at the genetic level, even though it has the same developmental pathway as one of the other vertebrates. The first geneticist concludes that homologous genes can control non-homologous developmental processes which bring about the formation of homologous structures. The second geneticist concludes that the genetics/development of the three vertebrates are not homologous, and therefore it is not possible to homologize the structures above that particular level. The problem arises with homologous processes, structures or phenotypes at one level; is it possible to make statements about homology at another level? Let me draw two conclusions about this. If one repeatedly found that homologous structures had different developmental pathways and different genetics, one could draw conclusions about the evolutionary process; the developmental pathways and their genetic bases have evolved. This should not be surprising because evolution is descent with modification and variation. If, on the other hand, one found that the same genetic basis was involved in generating the same developmental processes to make the homologous structures, one should be surprised. Such a constancy would indicate a constraint on evolutionary processes. In either case, one would be looking at aspects of evolution, i.e. descent with modification or constraint. In the general context of evolutionary theory one would expect to find variation at different levels rather than extreme conservation at different levels.

This example illustrates an important point that came out of this symposium, which is the importance of defining homologous *as* hands or *as* digits, etc. Elements are not homologues because they have a particular pattern of processes at a lower level. The challenge for the future is to find the ways in which connections are made between levels, i.e. between the gene level, the cellular level, the developmental organization level, the phenotype and the population behavioural level. How these convergent properties are translated from one level to another is the major biological problem we need to be concentrating on when we invoke the term homology.

Index of contributors

Non-participating co-authors are indicated by an asterisk. Entries in bold indicate papers; other entries refer to discussion contributions.

T

V

W

Subject index

D

E

H

I

L

M

N

O

P

Other Novartis Foundation Symposia:

No. 214 **Epigenetics**
Chair: A. Wolffe
1998 ISBN 0-471-97771-3

No. 219 **Gap junction-mediated intercellular signalling in health and disease**
Chair: N. B. Gilula
1999 ISBN 0-471-98259-8

No. 225 **Gramicidin and related ion channel-forming peptides**
Chair: B. Wallace
1999 ISBN 0-471-98846-4